THE NATURAL HISTORY MUSEUM OF LONDON BOOK OF

PREDATORS

THE NATURAL HISTORY MUSEUM OF LONDON BOOK OF

PREDATORS

HOW PREDATORS FIND, CATCH
AND CONSUME THIER PREY

THIS IS A CARLTON BOOK

This edition published by Carlton Books Limited 2001
20 Mortimer Street
London
W1T 3JW

ISBN 1 84222 400 X

Commissioning Editor: Claire Richardson
Senior Art Editor: Vicky Holmes
Art Direction: Peter Bailey
Picture Research: Sally Claxton
Production: Garry Lewis
Jacket Design: Alison Tutton

CONTENTS

1 PREDATOR WORLD

Out in the wild, one of the most feared warnings you can hear is: "Look out – a predator!" Is it a tiger, creeping up behind you? Or a pack of wolves, howling and snarling? A cobra with a lightning-fast strike and deadly venom? A shark closing in for the kill? Run – or swim – for your life!

Not necessarily. The predator could just as easily be a starfish gliding slowly over a seashore rock, a small green tree frog sitting on a leaf, or a dragonfly flitting over a pond. When we think of predators, we tend to imagine creatures that might prey on us. But the world of animals is vast and varied. Predators include not only lions on the African plains and eagles soaring over mountains, but little lizards sunbathing on stones, a swallow snapping up flies in flight, and worms in the mud at the bottom of the sea.

SO WHAT, EXACTLY, IS A PREDATOR?

As usual in biology, it is difficult to give an exact definition. In general, a predator is an animal that hunts. It feeds mainly by finding, catching and killing other animals – its prey. The predator eats the flesh of the prey. It might also consume other body parts, such as blood, bones and guts. Some animal groups contain nothing but predators. They include cats, seals, frogs and spiders. All members of these groups eat little else but the flesh or bodies of their victims.

Finding

A predator usually has sharp senses to locate its prey. On land, these senses include sight, hearing and smell. We have fairly good eyes and ears, so we can identify with a predator such as a hawk, which spots a small bird with its astonishingly sharp vision, and a fox, which hears the rustle of a rabbit in long grass.

In the sea, other senses are often more useful to a predator. They include senses to detect waterborne scents and smells, the ripples and currents set up as these animals move and swim, and tiny bursts of electricity that pulse out through the water from specialized electric organs. It is difficult for us to appreciate these types of senses, but they are vital to the daily life of aquatic predators.

FAR LEFT Growing to lengths of over 5 metres, the King cobra is the world's biggest poisonous snake – although it is not the largest predatory snake. Its bite can be fatal to humans.

BELOW Unlike the cobra, the red-eyed tree frog is not harmful to humans. Yet it is deadly to its prey – in this case, a cricket just within leaping distance.

Catching

Some predators creep up slowly on their victims; silent and stealthy. Others wait in ambush, perhaps hidden in undergrowth or rocks, or blended into the surroundings by amazing camouflage. Still others roam about in the search for prey, and when they find it they chase after it at full speed.

A predator's strategy may be in-built or instinctive, and quite rigid – that is, difficult for the predator to change. Predatory insects such as the praying mantis tend to have this "hunting by numbers" approach, since insect behaviour is largely instinctive. A predator such as a hyena, however, may learn to change and adapt its hunting strategies, perhaps for different prey, or at different times of year. Some of the most successful predators show this flexible, opportunistic catching behaviour.

BELOW The spotted hyena possesses, for its body size, the most powerful prey-catching equipment of almost any predator – its jaws and teeth.

This fat-tailed scorpion has gripped a cricket with its pedipalps (pincers), subdued it with venom from its telson (tail sting), and is now tearing apart the prey with its chelicerae (jaw-like mouthparts).

Killing

A predator usually has some specialized body part for killing its prey. For many, these weapons are strong and sharp – teeth, claws, beaks, pincers and the like. But the boa constrictor's weapon is its whole body, as it coils around a victim and gradually squeezes out its breath and life. The scorpion's weapon is the sharp sting at the end of its tail, which jabs powerful poison into the prey.

HOW ANIMALS FEED

Animals can be divided into four major groups, depending on the main types of foods they eat.

- Herbivores eat plants such as leaves, fruits, seeds and roots.
- Carnivores eat animals, including flesh and other body parts.
- Omnivores eat both plant and animal matter.
- Detritivores feed on dying, dead or rotting remains of once-living things, or their products. Earthworms and dung beetles are examples of detritivores.

A predator, therefore, is a carnivore. In theory, a predator could eat any kind of animal – herbivore, omnivore, detritivore, or even another predatory carnivore. But in reality, predators tend to avoid hunting each other. They would have to do battle with a victim that had sharp teeth, big claws, or other weapons perhaps resembling its own. Herbivores are usually an easier catch.

LEFT The grizzly is one type of the species known as the "brown bear". It is a massively muscular predator that could kill a moose. Yet it is a mere stripling compared to the greatest predator on Earth (*see right*).

Carnivores and predators

In general, a predator is a carnivore that actively hunts and catches prey which are relatively large compared to its own size. However, some take in large numbers of small prey; the blue whale, the largest carnivore on Earth, feeds by filtering sea water through the bristly, comb-like baleen plates in its mouth. It mainly consumes small, shrimp-like creatures known as krill. These are animals, so the blue whale is a type of carnivore. But the whale is vast, weighing 100 tonnes or more. Most krill are smaller than your little finger. Also the blue whale does not actively chase after or hunt down krill. It simply swims along, its vast mouth open, and the krill flood in. So, it is difficult to imagine the blue whale as a true predator in the way we think of sea hunters like killer whales and barracudas as predators.

The same idea applies to the biggest of all fish, the whale shark. It feeds in a similar way to the great whales, by filtering tiny

items from the plankton. But is it a true predator? Another type of shark, the great white, is generally regarded as the biggest predatory fish, even though it may be half the length and less than one-fifth the weight of a whale shark.

In fact, there is a whole range or spectrum of predation. Out-and-out hunters such as tigers and sharks take large victims. Hedgehogs and moles feast on more but smaller items. Hunter-scavengers such as hyenas and jackals are less fussy, and many bears and foxes are just as happy eating fruits and berries as fresh flesh.

LEFT Life and death struggles are not limited to the African savannah or Amazonian rainforest. They happen every night in the garden and park, as hedgehogs search for snails, slugs, grubs and other prey.

PLANET'S BIGGEST PREDATOR

What is the largest predator on Earth? A grizzly bear, a polar bear, or a tiger or lion? These are some of the biggest land predators. But look in the sea, and sizes increase dramatically. The great white shark is massive, at six to seven metres long and weighing a couple of tonnes. Even bigger is the killer whale or orca, at more than five tonnes.

But these are midgets compared to the mighty sperm whale. A big bull (male) sperm whale may reach 20 metres in length and weigh more than 60 tonnes. It dives deeper than almost any other air-breathing animal, more than 3,000 metres into the cold darkness. And it proves its champion-of-predators title by battling with one of the world's biggest prey, the giant squid – itself a huge predator of fish and smaller squid.

BELOW The sperm whale exposes its huge tail flukes, which measure up to 4 metres across, as it dives on another hunting trip into the permanently cold blackness of the ocean depths.

PREHISTORIC PREDATORS

The first sizeable animals appeared in the seas on Earth more than 600 million years ago. They were probably herbivores or detritivores rather than predators. Fossils show that there were seaweeds and other plants at the time, for them to eat. They were unlikely to be predators because, if they really were the first animals, there would be no other animals already there, for them to catch as prey.

But it was not long, in terms of the history of the Earth, before the first hunters swam in the ocean.

Fishy hunters

Fish appeared in the seas some 500 milion years ago. It took some time for huge predatory fish to develop. One of the most alarming was *Dinichthys*, also called *Dunkleosteus*, from the Devonian period, about 400–360 million years ago. At some nine metres in length, this monster was half as long again as today's great white shark. Its body was more eel-like than modern fish, and its jaws lacked true teeth, which had not yet appeared. But the jaws were edged with jagged blades of bone that could easily slice through its victims.

Sharks are perhaps the most infamous of modern predators, yet they are also a very ancient group. The first sharks slid through the oceans more than 350 million years ago. *Cladoselache* was around one metre long, with a slim, torpedo-shaped body, a large tail that could power it towards unsuspecting meals, and typical shark-like teeth to tear into the flesh. Sharks have changed little since.

BELOW Early amphibians may have hunted both in the water and, occasionally, on land. *Ichthyostega* was about one metre long and its fossils come from Greenland.

RIGHT *Tyrannosaurus rex* surveys its latest kill. Or has it just happened upon a dying victim by chance? For many years, discussion has raged about whether this great meat-eating dinosaur was an active predator or passive scavenger.

Predators on land

About 400 million years ago, life was spreading from seas and rivers on to the land. As in the water, the first land animals were probably herbivores or detritivores which ate the earliest land plants. But predatory centipedes, scorpions, spiders and insects such as dragonflies soon put in an appearance.

Meanwhile tetrapods – animals with a backbone and four legs – were also crawling on to dry ground, as their formerly fishy fins evolved into walking limbs. One of the first was *Ichthyostega*, a metre-long, swamp-dwelling fish eater from the Late Devonian period (360 million years ago). It was shaped more like a crocodile – a type of giant newt that swished its long tail to surge through the shallows and grab meals with its sharp teeth.

In pursuit of dinosaurs

Gradually, the first reptiles evolved from amphibian ancestors, and became more independent of the water. In the Triassic period, more than 220 million years ago, a subgroup of reptiles began to stride over the ground with a more upright stance, on their back legs rather than all fours. The stage was set for the biggest land predators that the world has ever seen. These were the dinosaurs.

Dangerous dinosaur predator

Perhaps the most famous of all dinosaurs, and certainly one of the most predatory, was *Tyrannosaurus rex*. It lived towards the end of the dinosaur era, in the Late Cretaceous period about 70–65 million years ago. Estimates of its size vary, but 12–13 metres long, five to six metres tall and six to seven tonnes in weight are average. Its biggest teeth, blade-like and with serrated edges, were nearly 20 centimetres in length – each probably bigger than your hand.

LEFT A model of the *Tyrannosaurus rex*, one of the largest hunter that ever walked the earth, showing its impressive set of teeth – the largest of which were nearly 20 centimetres long.

Back in the water

Dinosaurs never lived in water, but while they dominated the land, many other giant predatory reptiles terrorized the seas. The pliosaurs were especially fearsome. They had a long, crocodile-like head armed with sharp teeth, a streamlined body, four flipper-like limbs and a short tail tapering to a point. Two kinds, *Kronosaurus* and *Liopleurodon*, were simply huge – perhaps 20 metres long and more than 60 tonnes in weight. The skull of *Liopleurodon* alone was almost five metres long. The curiously thin, pointed fangs were longest at the front of the snout, where they grew to twice the length of the teeth of *Tyrannosaurus*.

Towards modern times

The dinosaurs and many other creatures died out in a mysterious mass extinction 65 million years ago. On land, their place was taken by mammals and large birds. These two animal groups soon evolved their own kinds of large predators. A strange hyena-like mammal known as *Andrewsarchus*, about four metres long, may have hunted or scavenged the bones of hoofed mammals resembling rhinos and tapirs.

ABOVE Many kinds of mammalian predators came and went over the past 60 million years. *Andrewsarchus* was the largest land-dwelling mammalian carnivore of all time, with a skull one metre long. Its fossil remains, found in Mongolia, date from 40 million years ago.

Over the past 50–40 million years, the predatory mammals familiar today such as cats, dogs, bears and weasels made their appearance. Sabre teeth were one terrifying weapon that appeared. Some 40 million years ago the leopard-like *Eusmilus* had them. So did the lion-like *Smilodon*, although this cat dates from less than one million years ago. The two long fangs in the upper jaw would have been efficient against large or dangerous prey. They were probably used to slash the victim so that it bled to death.

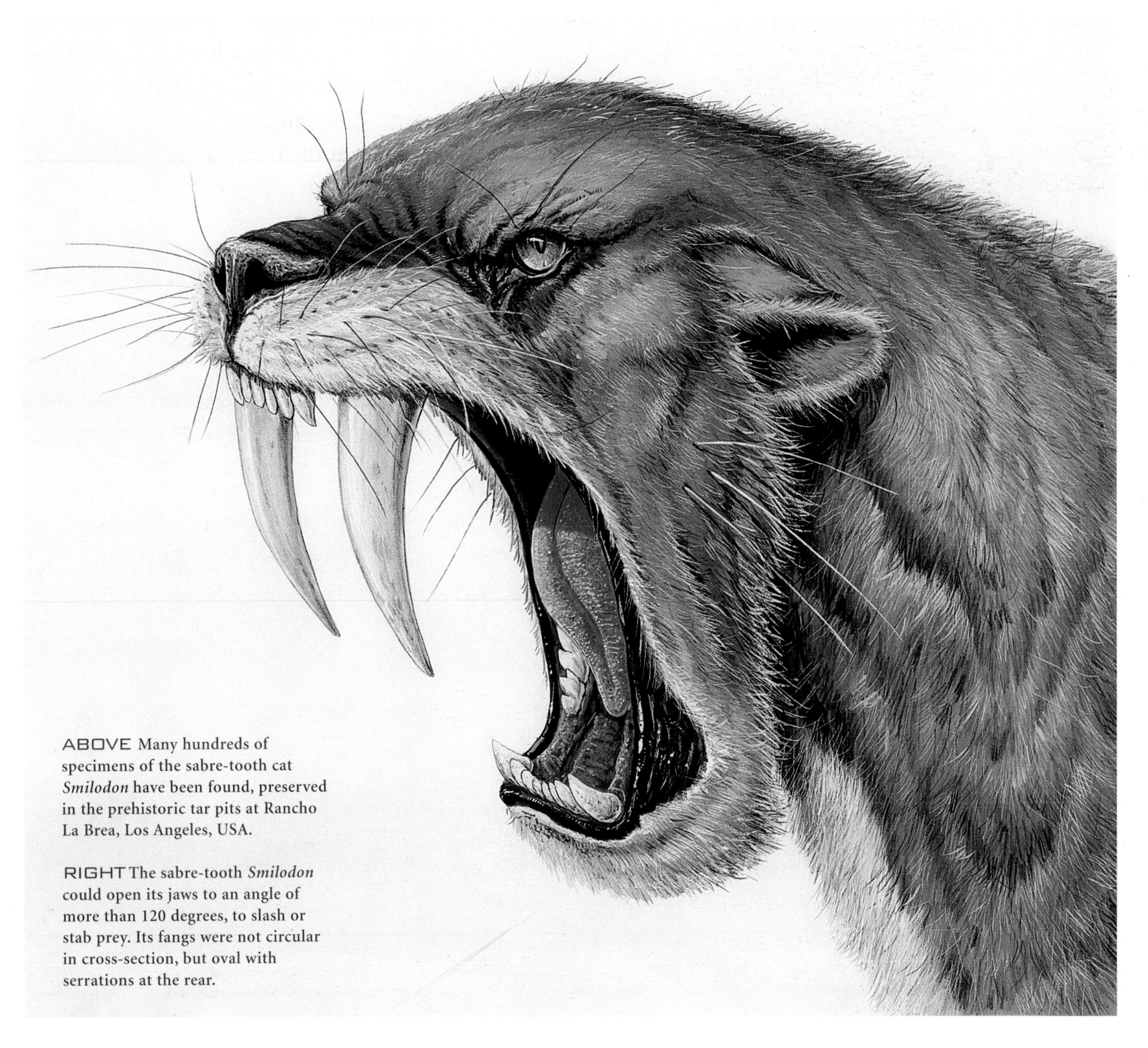

ABOVE Many hundreds of specimens of the sabre-tooth cat *Smilodon* have been found, preserved in the prehistoric tar pits at Rancho La Brea, Los Angeles, USA.

RIGHT The sabre-tooth *Smilodon* could open its jaws to an angle of more than 120 degrees, to slash or stab prey. Its fangs were not circular in cross-section, but oval with serrations at the rear.

BELOW AND RIGHT The edges of the puffin's beak come together like sharp-edged pliers to grip slippery prey. The beak itself is like an outer sheath made of horn, with the bones of the skull inside.

Predators or not?

Some animals are very deadly hunters in their own way, but they do not usually spring to mind when predators are mentioned.

Try these for size:

- The puffin is a dumpy seabird with an amusingly large, colourful beak. It does not seem fierce or frightening. But to small fish such as sand eels, it is death on wings. The puffin can carry more than 20 such prey in its beak.
- The jellyfish appears to be a harmless, innocent lump of hazy slime. Yet its trailing tentacles are armed with poisoned barbs that inject stinging, paralysing venom into fish, prawns and other victims (see pages 56–7). The food is gradually drawn into the jellyfish's stomach opening located in the middle of its underside.
- The plaice is a red-spotted flatfish that does not have predatory weapons such as huge claws or sharp fangs. Instead, it hunts over the sea bed for worms, shrimps, crabs and shellfish, which it crunches into pieces.
- The shrew is, for its size, one of the fiercest of all predators. A pygmy shrew weighing three grams will hurl itself, teeth gnashing and slashing, at spiky prey such as grasshoppers and cockroaches larger than itself. The shrew soon devours most of its victim. Then ... it is ready for another huge meal.
- The armadillo appears slow and inoffensive, as it shuffles and snuffles through the night, sniffing into holes and under logs. Yet it catches many prey, from ants and termites to lizards, birds and frogs.

RIGHT The nine-banded armadillo has large, powerful feet to dig its home burrow or den, and to unearth small animals as prey. This mammal has curved plates of horn protecting its body from larger predators.

Predators and parasites

Is there a difference between the hunting lifestyle of predators and the living-off-others lifestyle of parasites? In most cases, it is fairly easy to decide. But, as usual, there is a range of examples that shows the gradation from one to the other.

Consider the vampire bat. Is it a predator? All it ever consumes is blood. Under cover of darkness, it makes a slit in the skin of a large animal such as a horse or cow, with its razor sharp front teeth. Then it laps up the blood that oozes out. Full of the red liquid, it flies away. In most cases, the cut heals over and the victim suffers no lasting damage. On a much smaller scale, a mosquito does a similar blood-sucking job. These are parasites that feed off hosts, rather than predators which consume their prey.

How about the grubs or larvae of certain parasitic wasps? The adult female wasp searches for a caterpillar or similar host, as a living larder for her offspring. She stings the caterpillar to paralyse it, and buries it in a hole while laying her eggs on its body. The eggs hatch and the wasp grubs hungrily feast on the flesh of their juicy host. Finally there is just an empty skin, and the wasps leave to begin their adult lives. They have killed and consumed their victim, in the manner of a predator, yet they are called "parasitic" wasps.

ABOVE The false vampire bat is well named. It is not a "haemo-parasite" that sucks blood, like the real vampire bat. It is a true predator, and a large and powerful one – the biggest bat in the Americas, with a one-metre wingspan. It swoops on mice, rats, frogs, fish, birds and even other bats.

Hunting – an easy life?

Predators may seem to have it easy. Cats and dogs laze about most of the day, and just hunt now and again. Cuttlefish and starfish are rarely in a hurry. When these predators catch a meal, it is full of juicy flesh, goodness and nutrients. It is easy to digest and keeps them going for a long time.

In comparison, plant-eaters such as zebras, rabbits, sparrows and crickets appear to have a harder time. They are always on the go, searching for food. Their meals are often tough to chew, take a long time to digest, and, size for size, provide much less nutrition than meat. And they must always be on the lookout – for predators.

But being a hunter has its hazards. For a big cat like a cheetah or leopard, perhaps only one chase in ten might be successful. So there is a large amount of wasted time and effort. If the prey fights back, as it often does, the predator could suffer an injury that prevents future pursuit, or a wound that becomes infected. In harsh conditions such as drought or a big freeze, there may be hardly any prey at all. Likewise, in times of plenty, carnivores may breed so well that they become too numerous and go into "overkill", eating all the available food. These are just some of the dangers that ensure predators do not become too prevalent.

ABOVE A leopard snarls and bares its teeth to protect the kill it has just made. Carcass-robbing is a common problem among predators, and spotted hyenas are particular culprits.

BELOW Lionesses probably brought down this wildebeest after a few minutes of effort, but it will supply the pride, including this lion, with meat for five days. Predation is not always this easy.

KILLERS, BUT NOT PREDATORS

Were prehistoric humans predators? About 100,000 years ago, many massive mammals roamed the land, especially in northern regions of the world. They included woolly mammoths and their elephant-group relatives, the mastodons; also woolly rhinos, great bison, various types of horses, deer such as giant elk, and giant versions of kangaroos, ground sloths and armadillos. There were also huge predators such as cave bears, sabre-toothed cats and cave lions. Ice and snow gripped much of the far north.

By about 10,000 years ago, much of the ice had gone – and most of the large mammals had, too. Perhaps the rapidly changing climate had an effect, as natural global warming ended the Ice Age. But during the same period, humans were spreading around the world. We known that these ancient people killed big game with spears, axes, pit-traps, fire and other weapons. Fossil bones of the animals show cut and chop marks from butchering, and prehistoric cave paintings depict the hunting scenes.

So early people may have been responsible, at least partly, for this mega-mammal overkill. We do not know if the animals were slaughtered for meat and other body parts used as food, or maybe tools, or for their furry skins. Perhaps their bones, horns, teeth and claws were valued as hunting trophies and symbols of wealth and power. Possibly ancient beliefs, rituals and ceremonies played a part. The big carnivores could have been wiped out because they threatened people directly.

If so, this was the first great wave of human-caused extinctions. Sadly, the process continues at an increasing rate. So humans may not be out-and-out predators, in the sense of hunting for food. But we are certainly the most efficient killers that the world has seen.

BELOW A relatively new type of predator gathers in a pack to menace migrating mammoths and separate one from the herd. Armed with sharp spears and intelligent brains, these early humans were very likely to have obtained their quarry.

2 SENSITIVE KILLERS

Tradition teaches us about five senses: sight, hearing, smell, taste and touch. These are the most familiar to us, housed as we are in our human bodies. But nature in general, and predators in particular – especially aquatic predators – can do much better. They can double the list to at least ten. There are animal sensors specialized to detect infrared waves or "heat rays", electrical pulses, variations in the Earth's magnetic and gravitational fields, changes in the amounts or levels of natural salts and minerals in water (which gets saltier as a river nears the sea), and much more. Predators can also take our own five senses and extend their range into realms we can only imagine – spotting a rat from two kilometres away, or listening to sounds which are so high-pitched, even a dog does not notice them.

OPPOSITE The owl's eyesight is legendary. But its hearing is also acute, and it can locate prey on the darkest of nights, homing in for the final swoop almost by sound alone.

ABOVE The wolf spider spots prey using its eight beady eyes. Like its mammal namesake, the spider then runs down its victim. Courtship between male and female wolf spiders is also based on sight, rather than on scent, sound or touch.

AN EYE FOR THE KILL

Our own predominant sense is sight. Apparently, three-fifths of the knowledge in our brains comes in through our eyes. Indeed our own eyes are so highly developed that, compared to many creatures, human vision is almost a super-sense. So sight is perhaps the easiest of the predator's senses for us to comprehend – but a word of caution. Scientists can study the structure of animal eyes and measure their size, light-collecting abilities, focusing power and other features in immense detail. But it is far more difficult, and may even be impossible, to understand how the information from the eyes is analyzed and perceived in the animal's brain. How can we know what a creature sees in its "mind's eye"? This principle holds for other senses, too. We must beware of imagining that predators hear, smell or taste as we do.

Bigger is better

Two features of eyes identify the typical visually adept predator. One, they are big. In a daytime hunter like a wolf spider, frog, hawk or African wild dog, the eyes are dark, beady balls, prominent compared to the overall size of the face. In some smaller daylight predators, especially insects like the dragonfly and praying mantis, the eyes are bigger than the rest of the face and head together.

In night-active animals and those that hunt in the ocean depths the size relationship is even more startling. The major aim of such big eyes is fairly basic – to collect plenty of light, even on the most overcast, moon-less night, or in the deep sea, for detection inside the eye and analysis by the brain. The eyes of some small, tree-dwelling, nocturnal cats take up about one-quarter of the entire head. In many owls, the eyeballs are so large that they can hardly swivel in their sockets to direct the gaze up, down, or sideways. So the owl has to twist its neck to pivot its whole head around – some can actually peer directly behind themselves. The bulging eyes of a fish called the rough-headed grenadier, a type of deep-sea cod, are almost as big as its fins.

Judging distance

The second feature of vision-based predators, especially mammals, birds and invertebrates (animals without backbones), is that the eyes face mainly forwards. They are at the front of the head, not on the sides. There may be two eyes as in a cat or dog, or eight eyes as in the jumping spider, but they all look at the prey. This allows, among other benefits, accurate judging of distance so that the hunter can charge or pounce at the correct place. How?

Each eye sees an object such as the victim from a slightly different angle and so receives a different view. The brain compares the two views, especially in regions where the views from two eyes overlap. The more different the views are, the nearer the object. Also, each eye swivels inwards slightly to look directly at the prey. The angle of swivel increases as the prey comes nearer. We have this visual system and, from human-based knowledge, we understand that the brain can compare the angle between the eyes to judge the object's distance. This is known as binocular or stereoscopic vision. It is why we can judge distance much more easily when looking at a scene with two eyes through binoculars, than with one eye through a telescope.

BELOW The jumping spider's largest pairs of eyes help to gauge distance accurately, enabling the spider to leap with precision on its victim. Like all spiders, it trails a silk strand as a life-line, in case it should fall.

A cat, owl or lizard makes these binocular distance-judging features more sensitive by moving its whole head from side to side in a characteristic fashion. This increases precision in two ways. It provides a longer "baseline" ruler for eye positions and so differing views – something the jumping spider already has with its extra or peripheral eyes. It also produces parallax – the effect where nearer objects appear to pass in front of distant ones as the head moves from side to side. For example, it can tell a big cat predator whether its antelope prey is nearer or further away than a clump of trees. In nature's life-and-death struggle for survival, every little refinement helps.

Second-chance sight

Many nocturnal predators, especially cats and hyenas, have an extra refinement in the eye called the tapetum. This is a curved layer of reflective microscopic cells on the inner rear lining of the eyeball, beneath the main light-sensitive layer, the retina. Some light rays shine at the retina and by chance pass straight through, missing its own network of light-detecting microscopic cells. But the tapetum just behind reflects these rays like a bowl-shaped mirror, back along their incoming path. This gives the retinal cells another opportunity to trap them – and the eyes' owner is optically one up against its prey.

The rays which escape this second chance exit the eye along the pathways by which they came in. When a bright light shines into this type of eye, the tapetum causes a glowing reflection when seen from the outside. We call this eye shine and it is a familiar sight when we aim beams of light at nocturnal hunters. Also, although we do not have it ourselves, the tapetum has perhaps saved many human lives. It was the inspiration for inventing the cat's-eyes used on roads.

ABOVE Cat's eyes are designed for gathering the faintest rays on a dark night. In bright conditions the coloured, muscular iris enlarges to make the dark hole of the pupil into a narrow slit. This stops too much light entering the eyeball and damaging the delicate retina.

RIGHT The bald eagle's forward-facing eyes combine excellent binocular (stereoscopic) vision, to gauge distance, with a telephoto-like magnification to see items larger and in greater detail.

LEFT The fish eagle devouring a fish caught by spotting tiny clues on the water's surface, such as shadows, ripples and bubbles, and then plunging its talons into its fishy prey.

Detail of vision

Big, forward-facing eyes are a good start in the predator's sensory equipment. The eye's detailed structure is also tuned in many fascinating ways for detecting prey. In vertebrate animals (those with backbones), the light rays shining in are detected by a layer called the retina, lining the back of the inside of the eyeball. This has microscopic retinal cells which fire off nerve signals to the brain when light hits them. Like the tiny printed dots which make up the coloured pictures in this book, the more retinal cells there are, the greater the image's fine detail. We have about 125 million retinal cells in each eye, at an average of 200,000 per square millimetre. A bird of prey such as a buzzard boasts more than one million retinal cells per square millimetre, in theory giving five times the detail we can see.

Telephoto eyes

Another feature that sharpens the predator's vision for long-distance prey detection is the fovea. This is a patch of the retina where the light-sensitive cells are clustered closest together. When we look directly at an object, our eyes are angled so that its image shines directly on to the fovea, which is roughly circular in the human retina. For a cheetah on the African plains, the fovea is strip-shaped to give keenest vision where food is likely to be – along the horizon. Many birds of prey have both a strip fovea for scanning the horizon and a circular one to peer at a rabbit-shaped blob in greater detail.

Taking this feature yet further, the eagle's fovea is depressed – shaped like a bowl or shallow pit. This allows yet more retinal cells to be packed in. The result is that the central portion of the eagle's field of vision is magnified compared to the outer areas, by as much as twice life-size. It is like a built-in telephoto lens.

Bugs and bug detectors

A frog sees the tiniest fraction of movement in a nearby fly, judges direction and distance, flicks out its tongue and traps the victim on the sticky tongue tip – all in less than one-fifth of a second. But if the fly remains motionless, the frog pays no attention. Its vision may not put together a whole scene in the way we view the world. Each eye contains various types of retinal nerve cells, networked in groups to analyze certain aspects of the scene outside. Some cell groups react when they pick out moving edges or lines. Others, the "bug detectors", respond when small dots or specks pass by. This part-processed information is then passed from the eyes to the brain. The whole system is highly tuned to sensing prey and avoiding predators.

Seeing in the depths

Invertebrate hunters have a much greater range of eye structures and their eyes work in a more sophisticated way. The eyes of octopuses and cuttlefish are not that different from our own, and tests on octopuses show that they can detect detailed movements and also certain colours – another helpful ability for tracking down meals. The octopus eye has about 20 million retinal cells. But all these measurements are dwarfed by the octopus's massive cousin, the giant squid. Hunting through the dark ocean depths, it has the largest eyes of any predator, in fact of any animal. Each is 35 centimetres across and contains perhaps 1,000 million retinal cells. What the giant squid "sees" in its brain – who can tell?

BELOW A large damselfly would make a huge image in the frog's eyes. But the specialized "wiring" of the microscopic light-detecting cells inside the eye can also detect a gnat smaller than this "o".

Flight of the dragon

Insects have very different eyes to vertebrates, but they are still highly developed in the predators of the group. The insect eye is composed of many separate cone-shaped, six-sided units called ommatidia (see panel). The more ommatidia in an insect's eye, the greater its presumed ability to pick out fine detail.

At one end of the spectrum, underground bugs which have little use for their eyes may have as few as 10 ommatidia. Some praying mantises have several thousand. Champion is the dragonfly, which has an amazing 30,000. This is enough to spot midges, gnats and other likely prey, as the dragonfly hawks over its territory on the wing, or sits on a twig ready to dart out at passers-by. As a tiny victim moves across the field of vision, its dark image flicks from one group of ommatidia to the next.

The dragonfly's eyes also excel in another department – making out fast motion very accurately. It is able to distinguish individual wing beats of prey at the rate of up to 230 per second. Our own rate for this "flicker-fusion" effect is around 50, and much less in dim light. Above this, movements become blurs. This is why we see a bee's wings as a blur, but the dragonfly can follow the individual beats, perhaps as we follow a swan's individual wing flaps. Such high-speed, fine-detail vision allows the insect dragon to calculate the speed and direction of its future meal with astonishing precision, as it closes in for the kill.

ABOVE The hawker dragonfly's wings whirr as it is buffeted by the wind, but its eyes stay steady, locked on target.

As the chase takes place, the dragonfly's two main eyes are not distracted by another visual task, that of detecting the surroundings and horizon. Small, simple eyes on other parts of its head detect only light and dark, and this is enough to for the dragonfly to know what is horizontal and vertical, and which way up it is.

MANY EYES IN ONE

The insect eye is dome-shaped with a surface pattern of many tiny sections. Each of these is called an ommatidium. It receives light rays coming in from a small part of the surrounding scene, and detects their brightness or darkness and, in some cases, their colour. These small areas of visual information add together over the whole scene to produce a patchwork image, rather like a mosaic. However, it is not known whether the insect actually "sees" any type of mosaic image in its brain.

ABOVE The dragonfly's eyes dominate its head, the tiny units or ommatidia producing a rainbow effect.

Big ears for small sounds

A huge variety of predators in all kinds of habitats have big ears, so that they can listen for prey when all around is quiet. In the desert, the fierce daytime heat keeps most creatures in their burrows or sheltering under rocks or plants. Night is cooler and damper, and is the time to come out and eat. In North Africa, the fennec fox listens for the faint swish of sand grains disturbed by tiny feet, as gerbils, lizards and insects creep across the sand. Each of the fennec fox's ears is bigger than its face and can be bent and tilted to the best position for receiving the incoming sound waves. The pair work like twin radar dishes to pinpoint the direction of the prey. If the fox were the size of a person, each of its ears would be bigger than this opened book.

Such relatively giant hearing aids have a second use in the desert. They have many blood vessels inside them, and they help to get rid of excess heat from the fox's body, sending it into the surrounding air like cooling radiators.

ABOVE The fennec fox's huge ears act as sonar dishes to detect a sound source. In this front view the fox's left ear is turned to the side for this purpose.

CLICK-CLICK-CLICK-PREY!

As the bat flies, it sends out a series of sound pulses or clicks which are so high in pitch, most are ultrasonic – beyond the range of human hearing. (Young people can sometimes hear certain clicks, because their ears are more sensitive than those of older people to very high-pitched sounds.) These sonic bursts emerge from the bat's mouth or nostrils at the rate of 20 or more each second, and spread into the surrounding air. Any object in the way makes the clicks bounce back, or reflect, as echoes. The bat's huge funnel-like ears gather the returning pulses and track their volume and direction. Its brain analyzes the results with the speed of a modern computer. A "blip" on the sound radar could signify a moth or similar victim. Some bats are said to be able to find their way towards a gnat 20 metres away, by echolocation. They gradually increase the pitch and the rate of the pulses to more than 100 each second as they near the flying food. This provides more detailed information over a narrower area, helping them to home in for the final grab.

Sounds that come back

Bats find their way and their prey using their ears more than their eyes. Their method is based on making very high-pitched or ultrasonic clicks and squeaks, and listening for any echoes (see panel). This is called echolocation, and it resembles the sonar or echo-sounding systems of boats and submarines used for navigating at sea.

Bats are not the only hunting animals to use the method. In the ocean, dolphins also make echolocating clicks in their breathing passages, to assist their eyes in finding prey such as fish and squid. The noises emerge through the bulging forehead of the dolphin, which is called the melon and contains a fatty substance. The melon does the same to sound waves as a lens does to light rays. It concentrates or

focuses them into a narrower, more directional beam. As it sways its head around, the dolphin sends out beams of ultrasound, almost like people who explore caves examining their surroundings using a spotlight worn on a hard hat.

The dolphin is thought to detect the echoes by receiving them along its lower jaw, into the ear, which is mainly under its sleek skin. The timing and pattern of the reflections indicates the size and movement of the prey. It is not only dolphins in the sea that use echolocation – those in rivers do, too, and to even greater effect.

In the Amazon, Ganges, Chang Jiang and similar great waterway systems, the rare and little-known river dolphins swim in water so muddy that vision is often limited to less than one metre. Their eyes are tiny, but their ears and echolocation more than make up for poor vision as they hunt fish, shrimps, crayfish and similar food.

ABOVE A pipistrelle homes in on a caddisfly, guided by the invisible reflections of sound waves that are its own clicks. Some people can hear bats squeaking, but often these are not echolocating noises. Instead they are social calls, such as members of a communal roost identifying themselves, or a male attracting a female during his courtship flight.

OPPOSITE The barn owl's ears are small openings in its skull, well hidden by feathers. The dished, heart-shaped contours of the facial plumage may help to gather and funnel sound waves to each ear.

LEFT The owl's acute hearing does not only help it to hunt. The parent owl can hear the distress squawks of its chicks or the noise of an approaching, nest-robbing predator.

SOUNDS LIKE AN OWL

By day, hearing can help to guide a vision-led predator towards its prey. In darkness, hearing becomes the primary sense – even for some of the biggest-eyed hunters. Owls are famed for their ability to see on a moonlit night, almost as well as we see in daylight. Yet tests on barn owls show that they can catch prey such as a mouse even in total darkness – provided there are no distracting sounds. The answer lies not with the eyes, but with the ears. The owl's ears are as highly tuned to sound as its eyes are to light.

Like us, the owl has two ears for the same reason that it has two eyes. Binaural hearing is like binocular vision. Sound waves travelling through the air from one side reach the ear on that side slightly before they reach the other ear. The time difference is only a few thousandths of a second, but the ears and brain are sensitive enough to detect it. Sounds are also louder in the nearer ear. As the sound moves around to the front or back, the time and volume differences decrease. Directly in front or behind, the differences disappear. This is how we and listening predators such as the owl know which direction a noise is coming from.

Surround sound

The barn owl goes one better. Its ears are not level. The left ear is slightly high on the head, just above eye level. The ear opening beneath the feathers points slightly downwards. The right ear is slightly lower than eye level and its opening points slightly upwards. These offset ears give the owl additional time differences for sounds reaching it from above and below, which allow it to judge verticals as well as horizontals. In addition, the sound waves coming broadly from the front are thought to be collected and concentrated by the feathers on the face, which form a heart-shaped bowl, almost like a radar dish. They funnel the sound waves independently into each ear.

The whole system gives the owl "3D hearing", an auditory picture in three dimensions of where noises come from. As the bird swoops low over the sleeping countryside, it peers with its huge eyes, and listens for the merest rustle of a vole or mouse on the ground. But such ultra-sensitive all-round hearing can jam up on noisy nights, due to wind rustling leaves or passing traffic.

THE SCENT OF BLOOD

Nearly every creature gives off a smell, and some predators use this partly or exclusively to find prey. Even the familiar hedgehog, snuffling among leaves for insects and worms, locates victims chiefly by scent – its small snorts, making it sound like a miniature pig, gave it the name of hedge-hog.

Smell is based on the sensory detection of chemicals in the environment. Smell receptors are microscopic parts of the body specialized to detect small amounts of certain chemical substances. Different receptors are specialized for different chemicals. When the correct chemical touches its smell receptor, it makes the receptor fire off nerve signals to the brain, and the odour is perceived. Our own smell receptors are inside the nose. There are "only" about 25 million of them. They can distinguish thousands of different chemicals which waft in as we breathe. Our own sense of smell is not vital for survival, and gets distracted by the pleasurable scents and odours of perfumes, cooked food and air fresheners. So we are less aware of smell than of sight and hearing.

Out in the wild, smell is a much more serious business. Hunting mammals, birds and most reptiles smell in the same basic way that we do. But their noses are super-sensitively tuned into odours that say "food". The largest land predator, the polar bear, can sniff out a seal from ten or possibly twenty kilometres away. However, this is downwind and out on the ice, where there are few other scents to confuse the odour picture. In the forest, a grizzly bear or wolf can pick out the scent of deer from several kilometres away. The predator can then follow the invisible odour stream in the direction of increasing odour concentration, leading all the way up to the prey. On a smaller scale, many insects have chemical sense organs all over their bodies, especially on the heads, antennae (feelers) and legs, to track down prey by scent.

BELOW The scent of a wounded seal, stranded on an ice floe, carries for many kilometres in the clear, clean, chilly polar air.

THE SMELL TRAIL

Many land hunters sniff the ground. They are hoping to scent traces of the microscopic fragments of fur, skin, scales, feathers and other body surfaces that all animals leave behind as they travel. A mammal the size of a human sheds about 50,000 of these invisible flakes every minute. They drift down nearby like a invisible blizzard, and mark out the path we have taken, often for the next day or two, to any animal with a suitable nose.

We find such techniques difficult to imagine because our own smell receptors cover an area of only about five square centimetres, and because our meals are served to us on plates. A bear's or wolf's smell receptors can cover more than 150 square centimetres – one quarter the area of this page. Sniffing the air or ground makes the difference between a full meal or starvation.

Smell works at very short range too. In New Zealand, birds have evolved to live in a similar way to small mammals in the rest of the world. The New Zealand kiwi is unusual among birds, not only for being flightless, but for hunting mainly by smell. Its nostrils are at the end or tip of its very long, thin beak, rather than at the base. The kiwi forages and probes into leaves and soil, and both feels and sniffs for its prey of grubs, worms and other small soil creatures – doing the same job as hedgehogs elsewhere.

ABOVE Nostrils at the tip of the kiwi's extra-long bill sniff out soil-dwelling prey. While probing, a valve shuts off the airways at the base of the beak. After the kiwi has withdrawn its beak, it opens this valve by "blowing its nose", snorting hard out through its beak to clear the nostrils of clogging soil or sand.

Tasting smells

Snakes can see in the usual way, and some daytime pursuit specialists like whipsnakes and racers have excellent vision for chasing after their fast-moving prey. Many snakes also have good hearing, although they lack rounded patches on the sides of their heads that work as an eardrum, as in the lizards. They pick up vibrations in the air and through the ground via the mouth area, especially along the lower jaw.

Snakes also smell and taste for prey in the usual way – but have an added extra. The snake flicks out its tongue to wave in the air or dart down and touch the ground. The tongue is then pulled into the mouth, and its forked tip is poked into two small chambers in the roof of the mouth. Inside these are extra patches of chemical receptors, known as the Jacobson's or vomero-nasal organ. While outside, the tongue collects substances from the air and the ground. The tongue passes them to the chemical receptors in the organ for detection. The snake repeatedly flicks its tongue in and out, and uses this combined "smell and taste" sense to keep on the trail of its victim.

FIGHTING BACK

Prey can evade their predators by scenting the enemy. This happens very often, as the prey sniffs the air and then, even without looking around, flees the scene. It is a popular belief that predators such as big cats and hyenas work their way around so they are downwind and their own smell does not carry to their intended quarry. But hard scientific evidence for this is lacking. What probably happens is that a predator is more likely to scent a victim that is upwind than one which is downwind – and so is already downwind of its quarry as the hunt begins.

LEFT A grass snake may be non-venomous, but it is a deadly hunter of frogs, voles and other small creatures. Its tongue gathers microscopic scent particles both floating in the air and on grass, rocks, soil and other objects.

OPPOSITE The great white shark opens its jaws wide as it tries to grab the photographer's bait. The nostrils are small holes midway between the tip of the snout and the eye.

Smell in water

Odours come to land predators through the air, but for aquatic predators, they travel through the water. Hence water-dwelling animals' sense of smell is more similar to our sense of taste. Even so, a typical fish has two separate sets of sensory parts, nose-based smell and mouth-based taste.

The fish's nasal openings or nostrils lead into dead-end olfactory (smell) chambers, like elongated pockets. Muscular flaps guard the openings to allow water in and out as necessary. The linings of the chambers are folded to increase their surface area and, like the insides of our noses, contain millions of microscopic smell receptors.

In many fish, especially sharks, the front of the brain is just behind these chambers. A thick stalk of nerves carries signals from the lining of the olfactory chambers into the front part of the brain, for fast and detailed analysis.

The draw of the shark

Most predatory fish have a keen sense of smell. In particular, sharks have legendary powers of scenting the water for the merest hint of blood or body fluids from a likely victim. Up to one-third of the shark's entire brain may be devoted to analyzing the nerve signals coming in from its olfactory (smell) chambers. Tests with captive sharks show that some types can detect human blood at the strength of one part of blood in 100 million parts of water. This is the equivalent of one teaspoonful of blood in the volume of water that would fill an Olympic-sized swimming pool, or the remnants of a blood trail seeping from an injured victim 50 kilometres away.

In reality the ocean's swirling currents mingle thousands of waterborne chemicals, and the shark's sense of smell is less easily measured to parts per billion. But what is clear is that dangling a piece of bloody fish flesh just under the surface, in a region sharks frequent, is almost certain to attract them within minutes. For fish like the shark, smell is like long-range radar. It serves to detect prey from far away, and then other senses take over as the predator comes closer.

The deadly touch

For large land predators such as big cats and alligators, the sense of touch is vital when it comes to the mechanics of prey capture. But touch cannot really compete with sight, sound and smell to locate a faraway victim. However, there are myriad examples of the smaller killer being led by touch to find a meal. In a few cases, this is mainly by direct contact with the prey. Plovers, a type of wading bird, have bills which are more sensitive to direct contact than our fingers. The plover can find its tiny prey of worms, grubs, shellfish, shrimps and other seashore life in the mud by touch alone.

But in many more examples, feeling actually works at a distance. The star-nosed mole of North America has an array of 22 fleshy, whisker-shaped tentacles around its nose. They are ultra-sensitive to touch and help the mole to identify, among other things, writhing worms and grubs, as it tunnels through the earth. However, although it burrows through soil for shelter and nesting, this mole also spends a great amount of time hunting in water. An expert swimmer, it dives into ponds and streams and feels for small fish, aquatic insects, water worms, snails and similar food, especially in the mud at the bottom. It can pick up the wriggles of troubled victims by direct touch. Or it can sense them nearby by the ripples which their movements send out through the water.

OPPOSITE The copperhead or highland moccasin, is venomous, but less aggressive than many other pit vipers. Its heat-seeking pit organs are the dark areas between each eye and the tip of the snout.

BELOW Many kinds of plovers stride across beaches and mudflats, probing with their short, straight, touch-sensitive bills for worms, snails and other juicy items. These are silver-winged plovers.

HOT ON THE TRAIL

Emergency rescue services looking for people buried by an avalanche may use heat-seeking equipment, similar to the "night sights" of binoculars. These gadgets detect infrared radiation – waves which are similar in nature to light waves, but which are longer in wavelength, and so invisible to our eyes. The infrared waves carry heat energy and they can be "seen" by suitable sensors from a distance, forming a fuzzy image of sorts.

Predators have found a way to use infrared waves. The group of snakes called the pit vipers are so called because they have heat-sensitive pits or hollows, usually one on each side of the face, in the loreal region, which is between the eye and the nostril. The hole of the pit leads to a small but often deep chamber in the upper jawbone. It is lined with tiny infrared receptors which are ideal for picking up the heat from a nearby object – a warm-blooded mouse about 15 centimetres away, for example. Since the pit organs are paired, stronger heat signals on one side tell the snake that their prey is located to that side.

The pits are amazingly sensitive and can pick up differences in temperature of less than one-tenth of one degree Celsius. This helps to distinguish a warm mouse, sitting still and "frozen-with-fear", from the pebble next to it warmed by the sun. In total darkness, using its heat-seeking equipment alone, a pit viper such as a rattlesnake can pinpoint and strike accurately at a warm-blooded victim – mammal or bird.

The pit viper group includes cottonmouths (water moccasins), copperheads, bushmasters, sidewinders and other types of rattlesnake, the massasauga, the fer-de-lance and several other deadly biters which will crop up again in the next chapter. The boas and pythons, which are generally large, non-venomous, constricting (squeezing) snakes, also have pit organs, but these are usually sited on their upper lips.

Feeling from a distance

Extend this principle and the movements of likely prey, as they travel on their daily business, cause vibrations which carry for a distance in a medium such as air, soil or water. Numerous predators thrive on prey first sensed in this manner. The Malaysian cave spider can run through its familiar territory in total darkness, and sense the motion of insects and other potential prey by the air currents they send out as they scurry about. The spider holds up its front pair of legs, which function like an insect's antennae (body parts that spiders lack) and pick up the merest breath of moving air as it disturbs the fine hairs on these legs. This would not be possible out in the open, where breezes and winds would swamp the tiny airborne ripples. But in the constant stillness of the sheltered and pitch-black cave, the system is ideal.

Snakes are well known to pick up vibrations that are passing through the ground. After all, their whole bodies are in contact with it. The scorpion has a similar ability. On its underside are feather-like parts called pectines, which brush the surface and sense air- and ground-borne vibrations both from prey and heavy-footed enemies.

LEFT Barracuda shoal in large numbers around the edges of tropical reefs. The stripe-like lateral line along both sides of their body picks up ripples and currents, especially from fellow shoal members, as pressure differences in the water.

RIGHT A lobe-bodied argiope spider keeps in touch with its silken net by holding its legs, especially the second pair, lightly on the strands. Orb-web spiders usually reach maturity in autumn and so build their largest webs which are highlighted by dew droplets.

Signals along the wire

Spider silk is an ideal substance for carrying vibrations which are sensed using touch. The spider detects them by holding its silken strands with its legs and feeling the pulls and tugs of victims – just like an angler catching a fish by rod and line. Spiders have a huge variety of designs for their silk, including vibrating webs, tripwires, nets and ropes. These are described in the next chapter. Some orb-web spiders are so sensitive to touch, they can distinguish the struggles of prey from other debris which lands in their webs, such as wind-blown seeds.

Ripples and currents

Water adds a whole extra dimension to this sense of remote touch. Animal movements send out ripples and currents that, in calm conditions, can travel many metres through a lake or the sea. Fish, like sharks, have lateral lines, along the body to detect these ripples. The lateral line is a groove in the skin, or a tube just under it connected to the surface by pores (small holes). This groove or tube runs along the side of the body from the head region to the base of the tail and its inner lining contains microscopic cells.

Movements in the water ripple along the lateral line, sending signals to the brain. The system works in a similar way to an owl's hearing, and is able to distinguish front from back, side from side, and in many cases, up from down, since the lateral line is curved or arched along the side of the body. For example, a powerful ripple that begins near the rear of the body and passes forward, tells the fish that there has been a large movement behind it to that side. Many predatory fish, such as sharks, barracuda, marlin and tuna, use the lateral line to assist their other senses, as they draw nearer to their prey.

Shocking signals

The electro-sense is yet another sense in the array of detectors used by predators. Active muscles give off very weak pulses of electricity. Doctors can detect these on the skin of the human body and record them as a wavy trace on a screen. The heart muscle signals, for example, are known as an ECG, an electrocardiogram.

Such pulses hardly travel at all through air. But they pass easily through electricity-conducting water, and aquatic hunters such as sharks make great use of them. The shark's electro-sensors are in small hollows or pits called ampullae of Lorenzini, scattered in the skin of the head region. They work mainly at close range, a metre or two, although they become more useful in murky water. They pick up the electrical bursts or discharges from active, moving prey.

The hammerhead shark's extraordinary front end provides an extra-long distance between the eye, nostril and electro-sensors on one side and those on the other. Like most sharks, the hammerhead swings its head from side to side as it swims. The shark's eyes, taste receptors and electro-sense organs sweep the water, picking up sights, smells and signals which are stronger on one side than the other, and which could mean a meal is in the offing.

The shark's close cousin, the ray, has a similar system. As it lazily flaps and cruises just above the sea bed, the ray can detect electrical discharges from prey such as worms and shellfish, buried just out of sight under the mud.

BELOW The hammerhead's slab-like snout is thought to improve side-to-side sensory perception. It may also work like a hydrofoil as the shark swims, providing lift to buoy up its front end and so save energy.

Electric fish and duck-bills

Certain other groups of fish have a well-developed electro-sense. The Amazon knife fish detects prey with its own electricity, generated by specialized muscles at the base of its tail. These set up an electrical field in the water around the fish. Objects in the field, even if they are not producing electricity themselves, disrupt the field and so the fish can sense their presence. To use the system the knife fish needs to stay straight (hence its name). So it swims with its fins, unlike most other fish, which swim by bending their bodies. The mormyrid or elephant-snout fish of Africa have similar electrical abilities around their long, flexible snouts. The electric eel can turn up the voltage to shock, stun or kill prey.

One of the world's oddest animals, the Australian duck-billed platypus, is another keen user of the electro-sense. It is a mammal, but it lays eggs like a bird, and also has a flat, wide, leathery, bird-like beak or bill. The platypus dives into muddy creeks and billabongs (types of pools or lakes) to seek out prey such as freshwater shrimps, small fish, worms, shellfish and freshwater crayfish. The water is so murky that hunting by sight is usually impossible. But the platypus's beak, as it forages among plants and in the mud, is very sensitive to touch and to electrical pulses. It can detect changes down to millionths of one volt. This is enough to pick up the electrical bursts from the flicking tail of a small shrimp about one metre away.

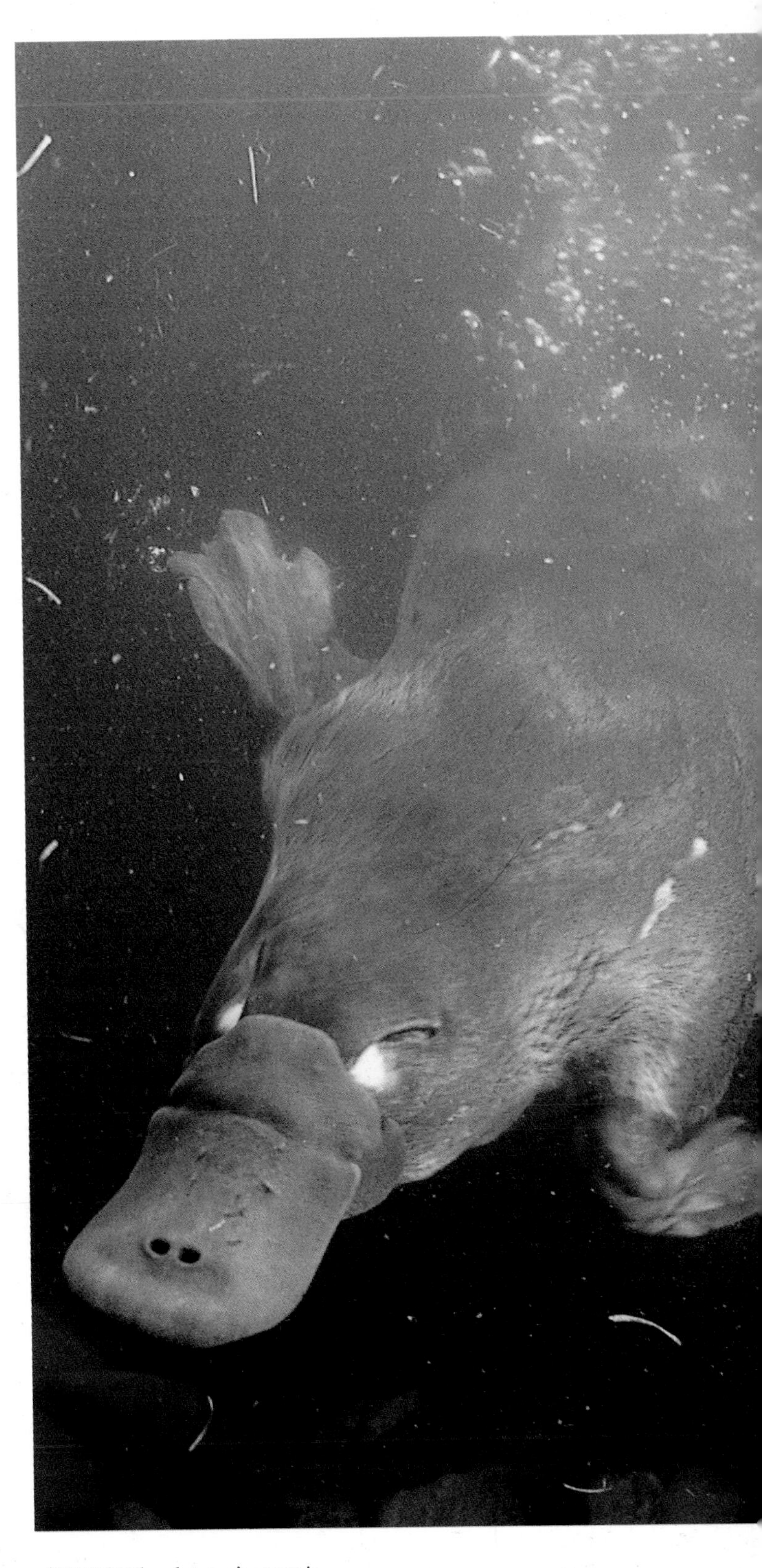

ABOVE The platypus's snout is covered by leathery, but very sensitive, skin. It detects touch, ripples in the water and electrical pulses given off by active animal muscles.

Find it – catch it

Most of the predatory senses described in this chapter have other uses. Bats employ echolocation, and knife fish use their electrical fields and lateral lines, to find their way as well as to detect prey. However, once the prey is located, tracked and approached, all predators face the next challenge: to catch it. Prey capture is the main topic of the next two chapters.

3 SNARES & POISONS

What is the most deadly poisonous animal in the world? One contender is the stonefish. This tubby, lumpy fish lies on the rocky sea bed in the shallows, around the shores of the Indian and Pacific Oceans. It has a truly astonishing resemblance in shape and colour to the stony, weedy sea bed on which it rests. The stonefish is a predator. It lurks camouflaged, watching with its upward-facing eyes for a passing shrimp, prawn or small fish. Then the stonefish rapidly lunges up off the bottom to grab the unwary prey in its gaping, upward-facing mouth.

Along the stonefish's back are thickened needle-like spines, which are parts of the dorsal fin. These normally lie flat on the fish's back but they can be tilted more vertical if an enemy approaches, or if the stonefish feels threatened. The spines are sharp, tough enough to penetrate beach shoes. The venom they jab through the skin is one of the most powerful in the natural world; every year several people die after accidentally treading on stonefish.

While the stonefish is dangerous to humans, prey capture does not actually involve the poison, which is kept purely for self-defence. This is an important distinction. The following pages deal with predators that do use their venom on victims, as well as those that build snares or traps to catch their meals. Two groups of animals excel at snaring and poisoning, and they are the creatures most people love to hate – spiders and snakes.

OPPOSITE Deadly venomous, but not for hunting – the poison fin spines on the stonefish's back are for defence. It feeds by simply gulping unwary victims into its wide mouth.

SPIDER SILK

No other animal spins silk quite like a spider. This amazing substance is stronger than almost any man-made fibre of similar thickness. The silk is made by glands at the rear of the spider's body. The glands have projections that resemble hollow fingers, known as spinnerets. The semi-liquid silk, a body protein, is squeezed out of holes in the tips of these fingers. As the substance emerges like toothpaste from a tube, the spider pulls and manipulates it with its rear legs. This pulling, coupled with the contact of the silk substance with air, makes the silk "set" into a flexible solid. A typical strand or filament of spider silk is only one-tenth of one-thousandth of one millimetre thick. The thicker lengths we see with the naked eye are several very thin strands lying together, produced from several adjacent spinnerets.

Spiders make different kinds of silk for different purposes. It can be thin and stretchy, thicker and stiffer, sticky or smooth, straight or crinkled or twisted. It is thought that hundreds of millions of years ago, early spiders evolved silk production as a method of wrapping up and protecting their eggs. Most still do this today, encasing the eggs in a silken cocoon. However, the silk has taken on extra jobs in many types of spiders. One of the most important and varied of these is prey capture.

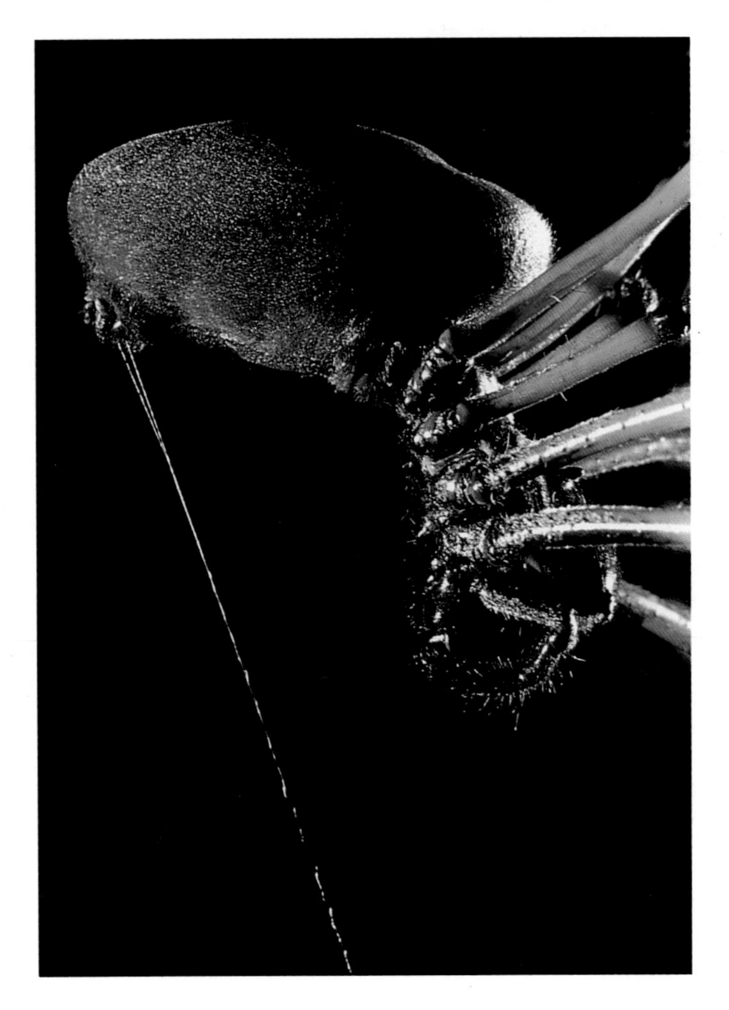

ABOVE Silk emerges as several ultra-fine filaments from the tips of spinnerets; finger-like projections at the rear end of the spider's body. Strands that are thick enough for us to see are usually several filaments together.

ABOVE A trapdoor spider cautiously opens the door to its home burrow, which is also lined with silk. The door is bevelled (slope-edged) so that it fits tightly into its frame.

Waiting behind the door

Different spiders spin and build an almost endless variety of silken structures to capture victims. The trap-door spider excavates a deep burrow, lines it with a mesh of silk threads, and makes a silken door, attached on one side by a small hinge-like flap, to fit the entrance perfectly. Here the spider waits by day, cool and protected. As darkness falls it partly opens the door and waits for prey to pass by. The unlucky meal is usually an insect of some kind like a beetle, ant, cricket or earwig. Most types of trap-door spiders dislike leaving their burrows completely, so they only attempt to grab victims that approach very close. Some types detect prey by feeling the vibrations it makes as it walks over the ground, using the sensitive hairs on their legs. Others in more open places peer with their beady black eyes from just under the partly ajar door. Still others use both sight and feel.

Once the trap-door spider has decided to attack, it lunges forwards and bites with its venomous fangs, holding the victim with its front pairs of legs. The spider usually tries to keep its rearmost legs in contact with the door and burrow. The time taken for the lunge and strike can be less than one-thirtieth of a second – not a lot of time for the prey to escape. As the venom takes effect the trap-door spider drags the meal into its lair for consumption.

In this example, the silk's main purpose is to form the camouflaging safety door for the burrow. Tripwire spiders use it in another way. As their name suggests, they lay silk strands along the ground near their burrows, often radiating outwards like the spokes of a wheel. Then the spider waits behind its door, front legs resting on these strands. Any insect that blunders into the trip lines tugs or twangs them and so is detected and seized.

Predator's purse

Purse-web spiders have taken the design a stage further. The silken meshwork lining their home burrow is angled and extended along the ground, or even part way up the base of a tree, like the tube-shaped hollow finger of a rubber glove. It is closed at the end and covered with bits of twigs, leaves and debris to blend with the surroundings. Inside, the spider waits. When it detects a small creature walking on the outside, it rushes to the same site on the inside, and spears the victim through the silk wall with its long fangs. The spider then cuts a slit in the tube, pulls the prey in, stores it safely in the depths of the burrow, and sews up the slit with more silk before commencing its feast.

BELOW Mesh-web spiders hide in a nest burrow, from which radiate many strands of silk. These are attached to a surface such as tree bark, rock or in this case an old stone wall. If a crawling insect like an ant or beetle disturbs the strands, the spider rushes out and bites it.

ABOVE A typical wheel-like orb web takes around an hour to construct. The spider usually eats the remnants of the old web, to recycle the silk's nutrient raw materials.

The orb-web

The most visible and striking snare that spiders spin to catch their prey is the orb-web. This incredible structure usually takes less than one hour to complete. It is almost invisible to our eyes, unless covered with dust or droplets of dew. Most orb-web spiders make a new web every day, eating the old one to recycle the nutrients. The web is a sticky net to trap prey which fly or walk into it.

First the spider selects a suitable space surrounded by twigs or other objects which can work as anchor points. In the upper part of the site, it spins out a silk strand and allows the breeze to carry this across, so that it catches on the other side. This top bridging line then becomes the main point for suspending more strands, which the spider attaches to more surrounding anchor points. The strands are attached, detached, moved, re-tensioned and reattached to produce an effect like the spokes of a wheel. Next the spider spins a non-sticky scaffolding spiral, starting at the centre and working outwards, round and round with one continuous strand, moving slowly from the centre to the outside edge. Then the spider reverses and spins a second spiral line from the outside inwards, laying it between the strands of the scaffold spiral. This second spiral is the sticky capture strand, covered with droplets of a glue-like substance.

Some orb-web spiders sit in the middle of the web, waiting for victims. Others hide behind a leaf or twig near the edge. The spider always keeps a leg or two on the silk, to detect the vibrations from struggling prey. It can feel the difference between a trapped creature and accidental debris such as wind-blown seeds. When a fly or moth hits the net, the spider races out, bites it with poison fangs, and quickly wraps it in another type of silk strand. The spider does not get caught in its own web because it walks on the dry scaffold spiral and avoids the sticky capture spiral. Also its feet have an oily covering that repels the glue-like droplets. Capture over, the spider takes its silk-wrapped prize to a shelter, for eating or storage.

A development of this throwing technique is used by the bolas and angler spiders. These are mostly fat, lumpy-looking creatures. Their technique is to spin one strong line of silk and place a sticky blob at the end, like a bead of glue. This is the "fishing line". The spider then holds or dangles the line, or whirls it around in a circle (bolas spider), and waits for a fly, moth or similar insect to flutter past. With amazing speed and accuracy, the blob on its line is hurled at the prey and sticks fast, as if to a patch of flypaper. The spider then reels in its food.

Even stranger is the hunting method of the spitting spider. It does not use ordinary silk, but a sticky, gum-like substance squirted from glands at the head end, linked to the fangs. As a victim passes, the spider lifts its head and shakes it from side to side, while squirting out two streams of gluey spit. This snakes through the air in a zigzag fashion and lands on the prey, tangling its legs and wings, and sticking it to the ground. Then, as usual for spiders, the spitting spider rushes up and delivers its venomous bite. For all spiders are poisonous – and this is the second main topic of these pages.

BELOW An antlion larva (young form) waits among the grains at the bottom of its pit-like sand trap. Its pincer-like mouthparts look more like horns on its head.

DOWN THE SLIPPERY SLOPE

The antlion is a type of insect, the adult of which is similar to a damselfly, although smaller and more delicate. It has a slim body, large eyes and long, lacy-veined wings. The young form or larva has a wide, plump body and no wings. It also has a massive pair of pincer-like mouthparts for grabbing prey. The antlion larva prefers places with soft, sandy soil. Here it digs a sand trap – a small cone-shaped pit. It hides just under the surface of the sand at the bottom of the pit, its pincers just showing. An unsuspecting ant or similar insect strolls past and begins to slip on the loose sand grains, sliding down the slope into the pit. The antlion larva bursts from hiding and grabs the victim in its great pincers. If the prey makes to escape, struggling up the slope, the antlion larva throws volleys of sand grains at it, and tries to knock it back down again. Once captured, the prey is quickly consumed by the voracious larva.

World of wonderful webs

The basic design of the orb-web has been adapted in many ways by other spiders. The triangle spider makes a three-sided web, while the stick spider spins just a single sticky strand about one metre long.

The web-throwing, web-casting, gladiator and ogre-faced spiders make small versions of an orb-web, which are so elastic that they contract into a patch just a few millimetres across, if they are not pulled out. The spider sits or hangs by its rear legs and holds the mini-web ready in its front four legs. As a prey walks or flies by, the spider widens its legs to stretch out the web, and places or throws it over the victim.

BELOW A day's wear and tear on the orb web can leave many holes and rips. Usually the spider will make a new web rather than attempting to repair the old one. Webs easily visible to our eyes have usually been abandoned and become covered with dust.

POISONS AND VENOMS

We can readily appreciate the physical weapons of a predator, such as teeth and claws. But often more deadly are the chemical weapons – poisons, venoms and toxins. Many widely different groups of animals have developed poisonous bites or stings for their predatory lifestyle.

All spiders can deliver a poisonous bite. This is usually done to subdue the prey. The poison comes out of a small hole near the tip of the tooth-like fang, which is part of the spider's main mouthparts, or chelicerae.

Some of the most powerful spider poisons belong to small species such as the closely related black widows of America, Europe and Asia, redbacks of Australia and katipos of New Zealand. They spin webs in the normal spider manner, then bite any trapped victims – not only insects, but even mice and lizards.

Given a choice as it strikes with its fangs, the spider often goes for the back of an insect's neck. In most insects, the main nerve (nerve cord) is quite close to the body surface as it passes through the narrow neck. The spider's venom mainly affects the nerves, and the muscles which these nerves control soon become paralysed. In the neck, the venom can reach the main nerve cord quickly and so work to disable the whole prey with the greatest speed and efficiency. The venom also contains digestive chemicals that dissolve the insides of the insect's body.

Some spiders have fangs strong enough, and venom powerful enough, to pierce human skin and cause harm to a person. See the Sydney funnelweb spider feature on page 126.

ABOVE Victim's view of a Sydney funnelweb spider's huge fangs being readied for a strike. If disturbed by a larger animal, this spider rears up the front part of its body and dribbles venom from its fangs.

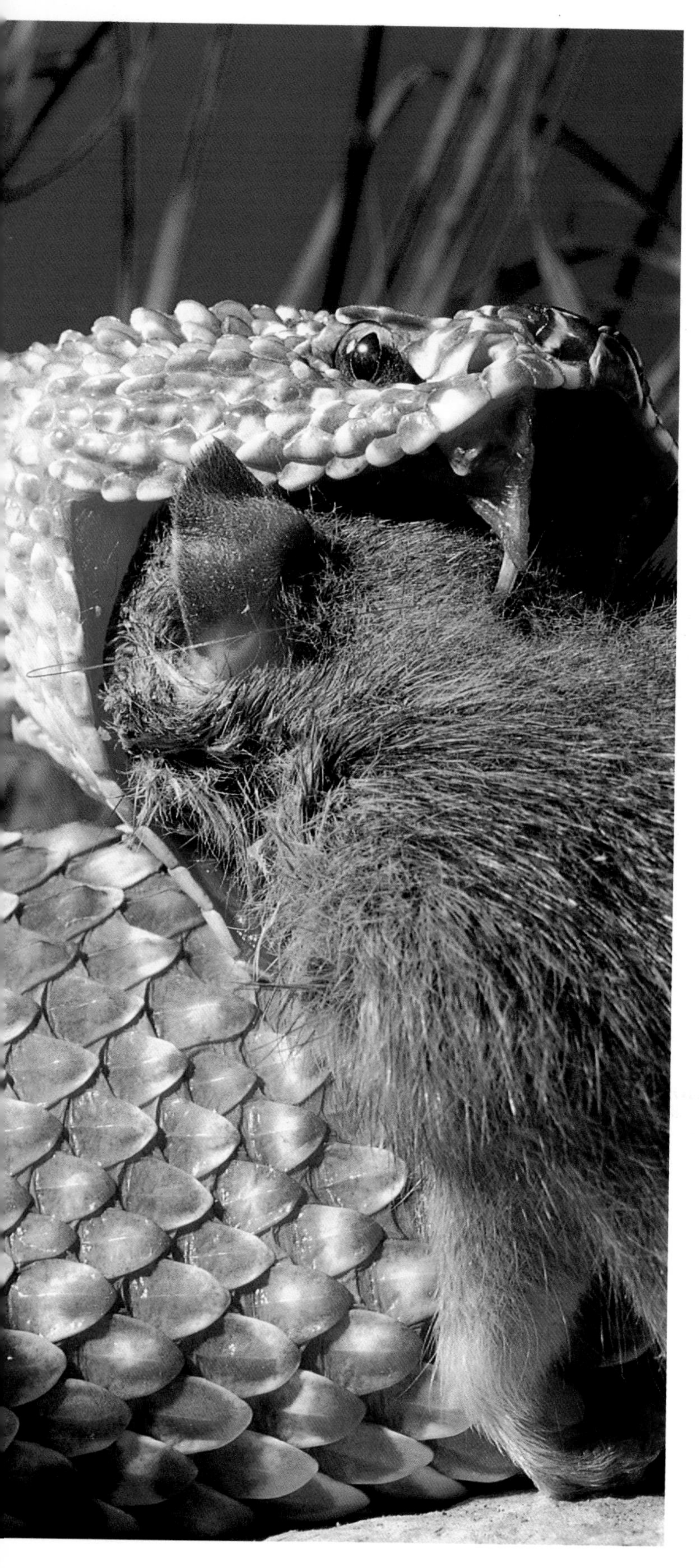

Strike to kill

Snakes strike fear into the hearts of most people. We may imagine they are all poisonous, even if their bites are completely harmless. In fact, relatively few snakes are venomous to humans – approximately 500 of approaching 3,000 snake species. Several groups of snakes have mild venoms which may be used in self-defence as well as in attack. Snake venom is a modified version of saliva, containing a cocktail of harmful chemicals that not only subdue prey but also begin to dissolve and digest its tissues. There are two main groups of venomous chemicals (see panel, page 52). The digestive ingredients include enzymes which quickly break down proteins, lipids (fats) and other substances that make up the body's microscopic cells.

Snakes lack the chewing molar teeth or the ripping canine teeth of mammal predators. The teeth of a poisonous snake are generally thin, sharp, backward-pointing and relatively fragile. They are adapted for a quick stab rather than a lengthy struggle. Many snakes use the tactic of striking fast, jabbing or injecting the venom with their longest teeth (fangs) into the prey. Then the snake holds on if the prey is not too large nor struggling too violently, as the venom takes effect. But if a snake, particularly a viper, fears for the safety of its fangs, it may withdraw to a safe distance. It tracks the victim, if this is still mobile, by scent and taste, while the poison works its damage. Once the prey has been disabled or inactivated by chemicals, the snake slowly swallows it whole.

LEFT A rattlesnake begins to swallow its victim after injecting poison through the tilted-down, clearly-visible fangs, one in each upper front corner of the mouth. Each fang is regularly replaced by a new one growing behind it and moving forward. If the two old fangs do not fall out on time, the snake may have four such teeth.

CHEWING IN THE POISON

The two main snake groups with large fangs and powerful venom, enough to paralyze or kill sizeable prey and even people, are the elapids and vipers. A few members of a third group, the colubrids, can also be dangerous to humans. Colubrids are sometimes known as rear-fanged or back-fanged snakes because their longer teeth are at the back of the mouth and each has one or more grooves along its length. These furrows allow the toxic saliva to flow into the prey's body as the snake works its meal to the back of its mouth and bites it repeatedly, almost with a chewing action. The dangerous colubrids live in Africa and include the twig snake and the boomslang.

ABOVE The coral snake is usually shy and secretive. Its bright colours warn other animals of its dangerous venom. It preys on other snakes and small lizards.

THE FRONT-FANGED SNAKES

The elapids, also called front-fanged snakes, include many of the feared types whose bites are fatal to people – mambas, kraits, coral snakes, death adders, taipans, the bandy-bandy and the many kinds of cobras. Their poisons tend to have major effects on the nervous system.

The elapids' fangs are two elongated teeth at the upper front corners of the mouth. They are fixed in position and may be grooved or have a hole running along the inside, like a hypodermic needle. As in other snakes the poison comes from venom glands, which are modified salivary glands in the sides of the face, below and behind the eyes. The snake opens its jaws wide as it strikes so that the fangs stab forwards and down into the victim.

TYPES OF POISONS

Snakes employ several main types of venom or poison. The venom of each type of snake usually has two or more ingredients. Different chemicals with similar effects are used by a wide variety of other venomous predators, from box jellyfish to scorpions. But in general they act on one of two major body systems in the prey.

■ Neurotoxins affect the nerve system, including the brain and spinal cord. Many types block the passage of tiny electrical signals through the network of nerves which spread through the body, and control the workings of the muscles. There are three possible fast-acting effects for the prey. The muscles in the chest used for breathing may become paralysed or go into spasm (uncontrolled, rigid contraction), leading to suffocation. The heart, its walls made of specialized cardiac muscle, may be affected in a similar way, and may stop beating (cardiac arrest). Or the skeletal muscles attached to the bones, which cause bodily movements, are paralysed. The victim cannot run away, or even move at all.

■ Haeomotoxins wreak their havoc on the blood and circulation. Some types, haemolysins, damage the microscopic red blood cells that carry oxygen around the body. This leads to localized bruising, bleeding and tissue damage due to oxygen starvation. Other haemotoxins are coagulants, which cause the blood to clot at the slightest opportunity. The clots build up in the blood vessels and block them, preventing life-giving oxygen from reaching vital parts such as the heart and brain. A third type of haemotoxin is the anticoagulant or thrombin, and it has the opposite result. It stops the blood clotting at all, especially at the bite wound. The victim bleeds to death.

FOLD-AWAY WEAPONS

The viper group includes the bushmaster, fer-de-lance, cottonmouth (water moccasin), copperhead, and the various vipers, adders and rattlesnakes. They have the most complex method of delivering their venom, which acts mainly on the victim's blood. A viper's fangs can be tilted by adjusting hinges in the jawbones. Normally the fangs lie flat against the upper jaw, along each side of the mouth. This allows them to be very long without projecting from the mouth when it is closed. It also keeps their points protected and sharp, since the fangs are stored in a fold or sheath of skin.

As a viper prepares to strike, it gapes its jaws and pivots the fangs forwards, so they swing down into the usual position, at right angles to the jaw. The snake also bends its neck to tilt its head back. The overall result is that the hollow fangs jab forwards with lighting speed. The venom is actively pumped through them as a lethal injection, and the fangs are then often pulled straight out again, to save them from damage.

BELOW A rattlesnake shakes its tail rattle mainly in defence, as a warning to enemies.

ABOVE A viper's fangs are usually folded back, along the upper jaw and roof of the mouth.

DEADLIEST OF ALL

Many venomous snakes prey on small rodents like mice, voles and rats, as well as small birds or bird chicks, little lizards, frogs, toads and occasionally fish, large insects, spiders or scorpions. But the world's largest poisonous snake – the king cobra of Asia, which exceeds five metres in length – eats other snakes, including venomous ones. Its fangs are not the biggest, only 15 millimetres in length. But its venom glands store enough poison to kill more than 10,000 of its usual victims – or about 150 people. The venom-spitting and hood-spreading habits of cobras are used mainly in defence, not when hunting.

BELOW The king cobra, the world's largest venomous snake can spread its "hood" by tilting the ribs in its neck region up and outwards. This is usually a warning that the snake may bite an enemy in self defence, rather than as a prelude to striking at prey.

Lizards and mammals

Among the vertebrates, only a few other groups apart from snakes use toxic or poisonous chemicals to poison their prey. The only two venomous lizards are the gila monster and the Mexican beaded lizard, which are close relatives and neighbours, both living in southern North America. They have grooved teeth in the lower jaw near the front of the mouth, which transport poison from nearby venom glands and prod it into the victim's wounds as the lizard chews. The venom helps to subdue the prey and predigest the meal.

Among mammals, only shrews have any kind of poison for attack. Their saliva contains a toxin which is extremely irritant and affects the nerve system of their prey.

SAFE FROM SELF-POISONING

Why are snakes, spiders, scorpions and similar poisonous predators not affected by their own venom, when they eat their prey? In most cases the venom must be in the blood or tissues to work effectively. It is already dispersed, "used up" and beginning to break down in the prey's body as the predator starts to feed. Then it is digested and broken down further by powerful acids and other chemicals in the predator's stomach. This digestive process destroys the venom's effects. It has been calculated that a rattlesnake would have to consume at least 100 times the amount of the venom it injects into a typical prey, to harm itself when it eats that prey.

The scorpion's sting

There are at least 600 kinds of scorpions, and possibly more than 1,200. But only a handful have stings venomous enough to kill a human. However, nearly all ready their stings in self-defence, and many use them to paralyse or disable prey. The sting is called the telson and is the last part or segment of the main body, the abdomen. It is almost full of venom gland. The scorpion strikes by holding its prey in its large pedipalps or pincers, and then jabbing its tail forward over its head, to bring the sting into play. The sting is rapidly rocked to and fro, using pairs of muscles attached to its base, so that its sharp, curved point seesaws and works its way into the prey's flesh. Meanwhile muscles around the venom gland also contract to force the nerve-acting poison, which may be as strong as that of a cobra, out of the sting's point and into its prey.

The centipede is another predator that employs poison to subdue its victims. Its venom-injecting, fang-like claws are actually a pair of front legs specially adapted for the purpose. (There are plenty of other legs just behind, for walking.) Large tropical centipedes grab and stab large insects such as locusts and cockroaches, and even lizards and mice. Their venom can cause great pain and massive swelling in humans, and even the rare fatality.

OPPOSITE Shrews are the only mammals with venomous bites. These tiny insectivores fight each other fiercely if one strays into another's territory.

BELOW In harsh habitats such as deserts, prey are rare. It's vital for predators to make sure of catching whatever passes by. This is one reason why many desert hunters, like this giant desert centipede, are so venomous.

ABOVE The beautiful colours and patterns of coneshells have long attracted human shell-collectors. But these molluscs have a jabbed-in sting which can cause terrible pain and suffering, and even death.

Fatal cones

Some of the most innocent-looking creatures of the sea are also the most poisonous. Coneshells are molluscs, cousins of whelks and snails. They live in many tropical seas, especially in shallow water – which makes them prone to being trodden on by people swimming and wading.

Different types of coneshells are slow but extremely efficient hunters of worms, sea slugs, shrimps and fish. The fish hunters have the most powerful poison and the surest method of delivering it. They have a long, flexible, finger-like part called a proboscis on the head, which has the mouth opening and mouthparts at its tip. The mouthparts include one or more hard "teeth", made of stony minerals, and shaped like hollow harpoons. A hungry coneshell near an unwary victim cautiously unsheathes and extends its proboscis, then suddenly jabs or fires the harpoon into its prey. A very strong nerve toxin is injected, that gets to work in seconds. The coneshell then creeps over its stilled meal, expanding its flexible body to swallow it whole.

Death by jelly

Even more innocent looking, but similarly poisonous and predatory, are the jellyfish and their relatives, collectively belonging to the animal group called cnidarians. The man o' war or bluebottle is a large jellyfish which is actually not a single creature, but a colony of individuals or polyps, living together and specialized for different tasks. The fishing polyps at the base of the colony trail their various tentacles for several metres through the water. When these make contact with a victim such as a shrimp or fish, poison barbs flick out from tiny stinging cells, or nematocysts, covering the tentacles. Hundreds of these barbs jab, hold, poison and paralyse the prey. The tentacles contract to haul it up towards the sucker-mouthed feeding polyps above, which share the digested proceeds around the colony.

Most jellyfish resemble the man o' war and snare prey with micro-barbed, stinging tentacles in a similar way. But they are single individuals, rather than floating colonies. Sea anemones and, on a smaller scale, coral animals (also called polyps) use the same prey catching method of sting-coated tentacles. They are right-way-up versions of the "upside-down" jellyfish.

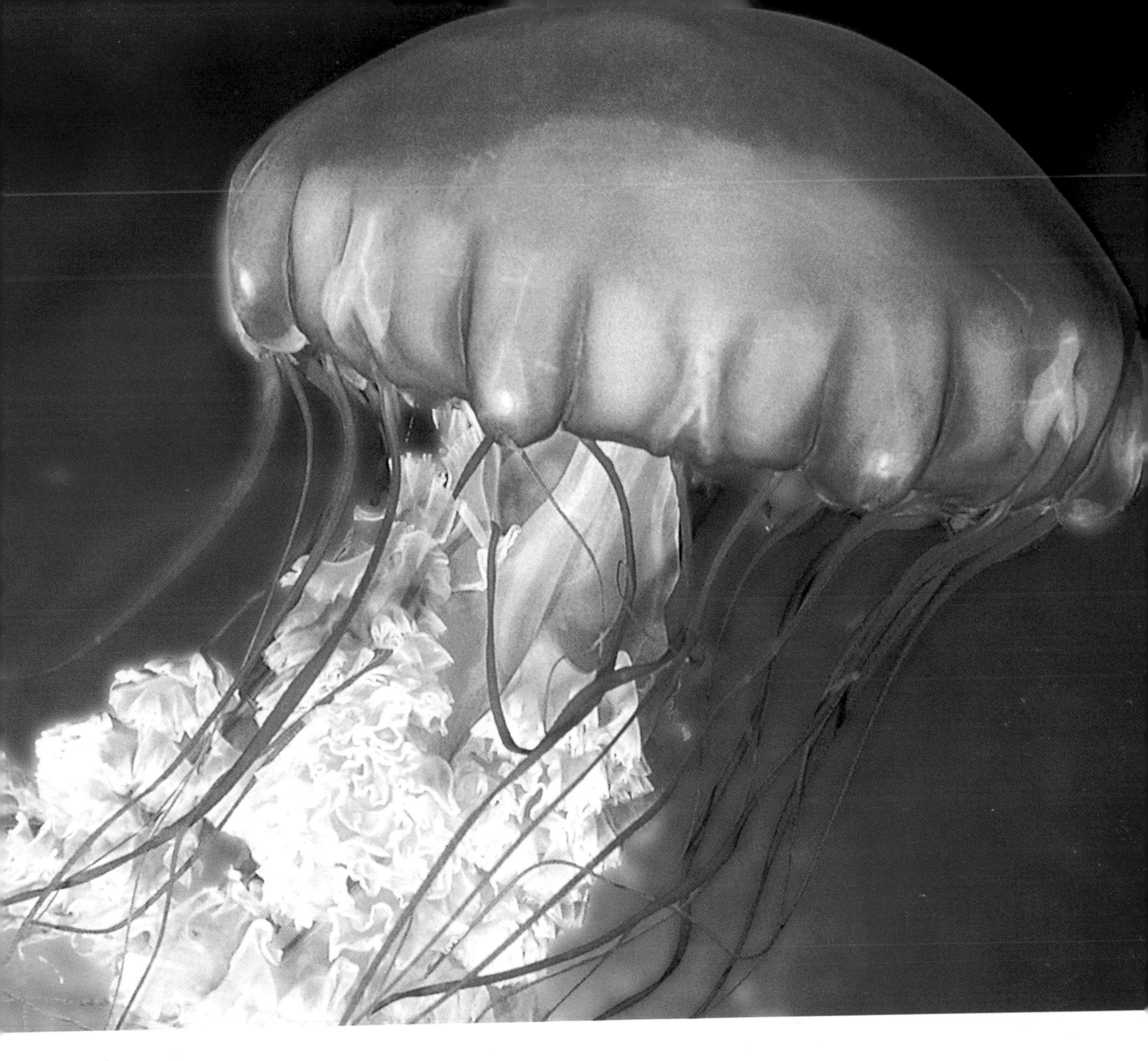

SEE-THROUGH KILLERS

Among the most feared types of jellyfish are the box jellies. They have angular box-shaped bodies but still trail their stinging tentacles. In fact the sliver-thin tentacles may be up to 10 metres long, attached to an almost transparent body 25 centimetres across. Millions of stinging cells on each tentacle contain one of the most powerful venoms known. People who brush against the tentacles suffer agonizing, burning pain, which has given these creatures the alternative name of sea wasps. Quite why such enormously powerful toxins are required to catch small prey like prawns and fish is not clear. Self-defence may be one key reason. The topic of animal predators that harm and kill people is continued in the last chapter of this book.

ABOVE Stinging tentacles trail from the edge of the sea-nettle's main body, the umbrella-shaped bell. The frilly parts hanging from the centre of the bell are mouth tentacles which gather and take in the prey.

4 WEAPONS OF WAR

The predator has located its potential meal. It has tracked its victim, or is in position for an ambush. Finally, the predator charges and seizes its prey. This sort of life or death situation occurs constantly in nature. However, if something goes wrong in the approach or charge phases, for example the prey realizes what is going on, the predator may break off rather than risk injury. The predator may pause and lie low, but if the prey has been spooked it is probably the end of the hunt. The predator can survive to hunt again later. The prey, once caught, cannot.

Predators have a fearsome array of physical weapons to secure their meals. These tools to catch and kill include teeth, claws and spikes. The weapons in this devastating arsenal are usually sharp-edged, pointed and penetrating. They jab, stab, tear and rip the prey.

THE IMPORTANCE OF TECHNIQUE

Prey capture is a highly evolved and coordinated process designed to maximize the chances of the predator taking and killing the victim, while minimizing the risks to the hunter.

For example, when a tiger charges it goes for a quick, clean kill. It uses its own body weight, and the strength of its massive shoulders and forelimbs, to knock or trip over the prey, toppling it on to its side or back. Almost before the prey can start to struggle back to its feet, caught by surprise and with the wind knocked out of it, the tiger has pinned it down and is at its throat. This is a matter of skilled technique, not random assault. In a split second the tiger has clamped its jaws on to the underside of the victim's throat or neck area. This is not the meatiest part of the body, but it is the most vulnerable. The windpipe, and the main blood vessels to the head and brain, pass through here, just below the skin. As the tiger's crushing jaws squeeze hard, the prey's breathing airway closes. Its brain becomes short of blood supply – suffocating and weakening, the victim's life rapidly ebbs away.

Yet still the tiger keeps its vice-like grip on the throat. It knows that some prey have a final trick: pretending to be dead. They appear to collapse, lose consciousness, cease breathing and go limp. The tiger loosens its hold, ready to start the meal. In a flash, the prey has kicked out, leapt to its feet and run away.

OPPOSITE After a stalk of possibly more than an hour, the tiger's kill may take just two minutes. Then it is time to move the carcass to a secluded place before feasting.

LEFT The tiger's array of teeth make short work of the prey. The long, pointed canine teeth rip and tear and the small incisors at the front nibble off meat. The shearing premolar and molar cheek teeth slice gristle and crack bone.

THE VARIETY OF TEETH

Like most cats, the tiger has 30 teeth. In each half (left and right) of each jaw (upper and lower) these comprise, from front to back:

- Three smallish incisors, used for nipping and nibbling – often when grooming as well as feeding.
- One huge, curved, cone-shaped canine, which is the main weapon for grabbing and stabbing prey, and for tearing and ripping off flesh.
- Three (upper jaw) or two (lower jaw) broad premolars. The third or rearmost upper premolar is a huge, long, sharp-edged tooth known as the carnassial. It closes together with the teeth below it like the blades of shears and can slice through leathery skin and snap gristle with ease.
- Finally, at the back, one large molar or cheek tooth for extra shearing and heavy-duty cutting.

A surprising extra "tool" of the typical cat is its tongue. This is very rough, coated with sharp-pointed, pimple-like structures called papillae. The tongue easily lacerates and softens flesh and rasps lumps from the carcass. It partly makes up for the tiger's lack of broad, flat-topped, chewing cheek teeth.

LEFT Anyone licked by a pet tabby knows that a cat's tongue is rough, almost like sandpaper. The papillae are small, sharp projections on the tongue that can rub, soften and loosen meat from bone. The rasp-like tongue is also used when grooming, to dislodge skin parasites such as fleas, and to comb tangles from fur.

Large and small

This pattern of teeth – incisors, canines, premolars and molars – is repeated in most groups of predatory land mammals. These mammals make up the group known as the Carnivora (see box p.63). The bears, and big cats like the tiger and lion, are the largest members of the Carnivora.

Remove the skin and flesh from a predator's head, and the skull, jaws and teeth are revealed as perfectly designed killing weapons. Like the tiger, the lion has immensely powerful jaws to crush its prey's windpipe, suffocating it. These predatory weapons of teeth and claws are found throughout the Carnivora. The weasel is the smallest member of the group. While its skull is smaller than an adult human's thumb, it shares many characteristics with a lion's skull, despite the difference in size. Although the weasel is one of the smaller mammal predators, it can kill a rabbit with its powerful jaws and tough, sharp teeth.

ABOVE The weasel's long, slim, flexible shape is ideal for following mice, voles, lemmings and similar prey into their burrows. It also hunts birds, reptiles and insects, and sniffs out nests for their eggs.

BELOW The weasel's skull may be tiny, elongated and flattened, yet it shows similar dental design to the skull of a big cat or wolf – long, pointed, fang-like canines and shearing, slicing cheek teeth.

NOT JUST A SCAVENGER

The hyena, another member of the Carnivora, is a misunderstood predator. Its skull is stub-nosed but hugely strong, with massive jaw muscles and enormous cheek teeth. The hyena can deliver an immensely strong bite, and even crack open bones to eat the nutritious marrow inside. This ability has led to the hyena's reputation as a scavenger, living off the kills of others.

But hyenas are also active hunters. Like cats, their hunting technique depends on creeping up on their prey slowly, then giving a short chase before killing with a single, suffocating bite to the neck. And hyenas are not always lone predators. They live and hunt in packs to take live prey as large as zebra and wildebeest. Indeed, lions and hyenas compete for the same food on the African plains, and hyenas have been seen to drive a lion away from its kill (see page 107).

ABOVE The spotted hyena pack has brought down a topi – a type of antelope standing 125 centimetres high at the shoulder and weighing more than 150 kilograms, with long, sharp horns.

WHEN PREDATORS ARE NOT CARNIVORES

Most predators are, by definition, meat-eaters or carnivores (with a small "c"). However, not all predators are Carnivores (with a capital "C"). The Carnivora is a group of mammals, most of which have a hunting lifestyle, and many of which have specialized teeth called carnassials (see p. 60). Members of the Carnivora include:

- Canids – wolves, dogs, foxes and jackals.
- Ursids – bears.
- Procyonids – racoons and the similarly-shaped coatis and kinkajous.
- Mustelids – weasels, stoats, otters and similar long-bodied, short-legged hunters.
- Viverrids – mongooses, meerkats and the cat-like genets, linsangs and civets.
- Hyaenids – hyenas.
- Felids – cats, big and small.

The Carnivora is part of the traditional grouping or classification of mammals and other animals. The system has been organized according to features of body structure (anatomy), especially jaws and teeth. However, it has led to various puzzles. From its body structure and also its genes (genetic material or DNA), the giant panda is often regarded as a member of the bear group, and therefore one of the Carnivora. Yet it is almost entirely vegetarian. Similarly, other mammals which are out-and-out carnivores, such as seals and dolphins, are not classified as members of the Carnivora.

RIGHT Meerkats catch and eat a huge variety of prey, from small birds, lizards and mice, to insects, spiders, snails and worms. And, like many smaller "carnivores", they also forage for roots, shoots, fruits and other plant matter.

THE SMILING SHARK

Unlike land predators such as big cats and bears, which have a variety of tooth shapes and claw-equipped paws, the ocean's ultimate killers – sharks – have only one style of weapon. This is their teeth. A shark's teeth are giant versions of the tiny, pointed denticles, or placoid scales, that cover its body and are embedded in the skin. Around the edges of the jaws these scales have evolved much greater size and strength, and are covered by a hard, enamel-like substance that resists wear. Under the enamel is dentine – the same tough substance that forms the bulk of our own teeth.

The shape of the tooth varies from shark to shark, and there may even be slight design variations in different positions within the mouth of the same shark. In general, however, each tooth has a surprisingly thin, triangular-shaped crown (exposed part), like the tip of a sword's blade, with two roots securing it into the gum and jaw.

The great white has 40–60 of these razor sharp teeth in use at any time, in the upper and lower jaws (see opposite). The largest of these are up to eight centimetres tall and about the same measurement across the base. Sharks such as the sand tiger have taller but less broad teeth, often with a main peak flanked by two smaller points, resembling a miniature mountain range.

ABOVE The sand tiger shark's teeth are almost T-shaped, with a wide horizontal base that tapers rapidly into a long, slim, sharp, fang. These teeth are suited to gripping small fish. Their wide spacing along the jaw gives the shark its alternative name of ragged-tooth shark.

When the shark bites

The two exposed edges of the triangular tooth have small saw-like teeth or serrations, making the tooth a very effective cutting tool. The shark clamps its teeth on to the victim and then shakes its head from side to side, so that the teeth saw through flesh, gristle and even bone. In many sharks the teeth are angled backwards from the gum, sloping into the mouth. This makes it even harder for prey to struggle free.

Studies to measure the power of the bite show that a big shark can exert a force of three tonnes per square centimetre. This is about the same as the weight of an elephant concentrated on to the area of a postage stamp. No wonder it can shear a seal in half or carve a massive lump of flesh from a whale.

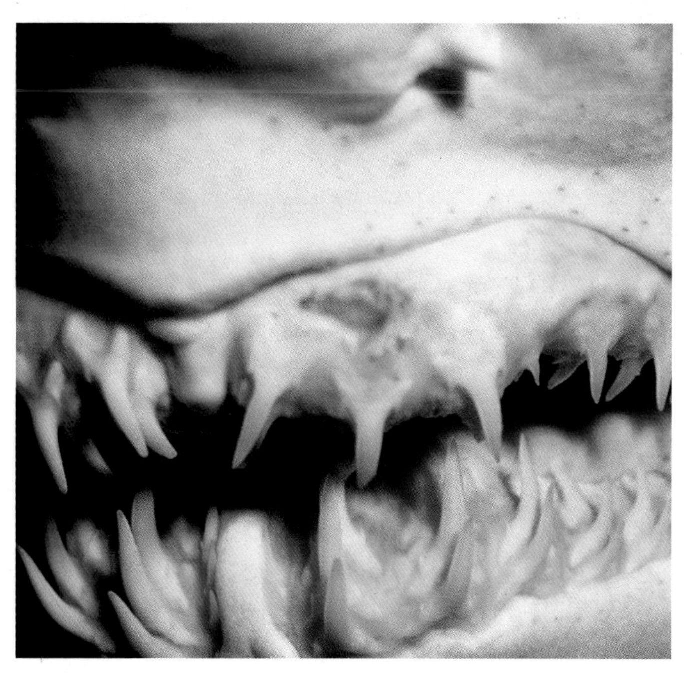

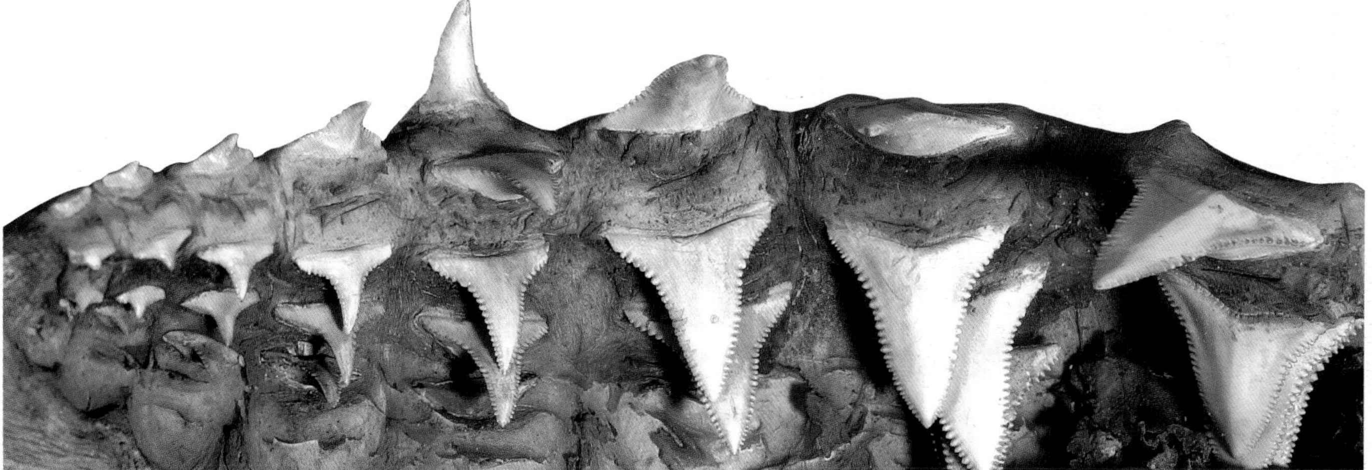

New teeth for old

The teeth we are able see from the outside are only a few of the shark's total dental complement. Each tooth begins its career as a small bud or pimple further within the mouth, on the inside of the jaw. Gradually it grows larger and also moves forwards, towards the rim of the jaw. After a journey lasting a month or two, the tooth moves forward up on to the rim of the jaw and into its working position.

This tooth growth and movement happens all the way along both jaws. Rows of teeth develop and pass from within the mouth to the edges of the jaws, where they may be seen as they come into use. The teeth at the front wear away and are shed, or occasionally snap off under stress. In general, one tooth is lost every few days. This is why many sharks have misaligned rows or missing teeth. However, there are always replacements coming from behind. A great white shark may get through thousands of teeth in a long lifetime.

TOP AND ABOVE New teeth are always developing in a shark's mouth, just behind the teeth in use. They usually lie flat, out of harm's way, as they slowly move forwards, as though on a conveyor belt. When they reach the rim of the jaw they tilt around to project vertically into use.

BIG BITE, NO CHEW

Only mammals have teeth that differ greatly in shape and use, such as incisors, canines and molars, within the same set of jaws. Most other predators have to make do, like the sharks, with teeth that are all much the same. A fine example of this is the crocodile.

The crocodile's teeth are fairly short, peg-like, not especially sharp, and well spaced along the jaw. They are specifically designed for grabbing and tearing flesh, rather than for chewing it. This action is supported by the structure of the jaws and jaw muscles. The mouth can close and clamp on to meat with huge power, but the jaw muscles used for opening the mouth or carrying out chewing actions are relatively weak.

The crocodile or alligator gets around this inability to slice off pieces of food and chew them, with its capture technique. It seizes and holds the prey underwater, then rolls or spins on its own length like a top, to twist and rip the victim apart. Large prey often die from a broken spine as the crocodile rolls over and over. Then the predator loosens its grip, grabs another portion of the carcass such as a limb, and does the same, rolling and twisting. Gradually the prey is ripped into pieces and the crocodile or alligator can swallow it as bite-sized chunks.

ABOVE Crocodiles and alligators have teeth which are all much the same in shape – conical. Each tooth is replaced every 6–24 months. Individual growth rates and replacement times differ, which is why the teeth are all different sizes.

FISH-EATING FEATURES

Teeth like those of the crocodile, which are relatively short, pointed or conical, also numerous, regularly spaced along the jaw, but not tremendously pin-sharp or blade-like, crop up in many kinds of aquatic predators. They include the cetaceans – porpoises, dolphins and toothed whales. This tooth design is often termed a piscivorous (fish-eating) dentition. It is suited to grabbing slippery, wriggling prey, especially fish and squid, and flipping these around into a suitable position to be swallowed whole. There is little or no chewing.

The types of seals and sea lions that eat mainly fish, rather than those which take a mixed menu including sea birds and shellfish, also tend towards a piscivorous dentition. Their teeth are more similar in size and shape along the jaw than the usual differentiated teeth of mammals. The same applies to the fish-pursuing otters. Apart from their large canines, otters have premolars and molars that are more similar in shape than those of closely related land predators such as the stoat or wolverine.

BELOW A bottle-nosed dolphin surveys a shoal of fish ready to surge forward, snapping with the hundred or more small, conical teeth in its long, beak-like mouth.

FANGS

Fangs are a specialized type of long, thin, sharp teeth. Snake fangs are described in an earlier chapter (see page 51). Fangs are used to stab and impale prey and are invariably angled or backward-sloping towards the interior of the mouth and the gullet. This helps to restrain the struggling prey, preventing its escape.

Some of the most remarkable fangs belong to the viperfish of the deep sea. They are almost as long as the fish's head. The viperfish can open its mouth extremely wide, enabling the fang tips to move far enough apart to be closed on to the prey. But the feeding methods of these mysterious eel-like fish, which dwell in almost permanent blackness, are not well known. The fangs may also function as the bars of a cage, to stop small prey from swimming off as the mouth closes around them. These small prey are attracted by a light lure – a glowing blob of flesh at the end of a long "fishing rod" fin spine. The spine extends from the viperfish's dorsal fin on its back, over the head, to dangle the lure just in front of the deadly fangs.

ABOVE Deep-sea predators like the anglerfish (pictured) and viperfish have needle-like fangs, to stab any prey that comes within reach. Substantial meals are rare at the bottom of the sea.

ABOVE RIGHT Like many raptors (birds of prey), the golden eagle holds down the prey – here a cottontail rabbit – with its feet while tearing off meaty strips using its beak.

THE BILL AS A WEAPON

Birds lack teeth, which are heavy and would slow down a lightweight flyer. Instead, the bird has a bill or beak, made of a very adaptable and light but strong horny substance. The horny bill covers the underlying jaw bones, or mandibles.

A bill is highly specialized for its owner's way of life. The size and shape of the bill are closely linked to how a predatory bird captures and eats its prey. For example, the hooked bills of eagles, hawks and owls are ideal for tearing flesh and dismembering prey. The bill does not usually seize and kill the prey – the long, sharp, curved talons (foot claws) are the weapons that do that. Large eagles such as the harpy, monkey-eating and golden eagles have fearsome taloned feet which are as large as human hands, and with an even stronger vice-like grip. The talons impale or crush the victim as the bird flies back to a feeding post such as a favoured branch or crag. Here it can release the prey and begin to feed.

The hooked bill of a bird of prey works rather like a combination of knife and meat hook. As an owl begins to feed, the highly curved, tapering upper part of the bill, which curls over the lower portion, forms a hook that digs into the meal. The bird jerks its head backwards to rake the hook through the flesh and rip off small pieces to swallow.

Jab and stab

The heron has an entirely different design – a long, dagger-like bill for spearing fish, frogs and other aquatic snacks. The exact method of attack varies with the circumstances. It may be a sudden grab of the prey between the mandibles. Or a simple jab or stab with the closed bill, to wound the victim, making it easy to pick up soon afterwards. Another method is to gape the bill slightly and prod with it in the hope of impaling the meal on one or both points.

There are dozens of other examples showing how bird bills are allied to a predatory lifestyle. The albatross, which has the longest wings of any bird, spanning three metres or more, also has a very long bill with a downturned tip on the upper mandible that slots over the curved end of the lower mandible. This shape is ideal for snatching food items such as fish, squid and swimming shellfish from the surface of the sea. Food can also be held in the bill until it is convenient for the albatross to swallow it.

BELOW Albatrosses soar for months over the open ocean, looking for prey on or just below the surface of the water. The meal is snatched during a swoop, which uses less energy than landing and then having to take off again.

ABOVE The unique beak of the avocet turns up at the tip. When the bird bends down to feed by dipping the bill into the water, the tip lies practically horizontal allowing the avocet to feed in very shallow pools, and on the mud or sand left by the retreating tide.

More birds and bills

Even longer, at about 60 centimetres, and more capacious than any other bill, is the bill of the pelican. It has an expandable bag of tough, stretchy skin below it, in the chin area, called the gular pouch. This is not a long-term storage sack for fish and other food, but an aid to their capture. Some types of pelicans feed while swimming or paddling at the surface while others dive-bomb into the water from several metres above the surface. Usually, the bird sweeps its bill through the water and the pouch fills like an expanding balloon. Then muscles in the mouth, throat and neck area make the pouch contract, spilling water out through the narrow gap between the upper and lower mandibles, rather like pouring liquid from a jug with a lid. Food items are trapped inside and rapidly swallowed.

One of the most unusual bird bills is the avocet's. This striking black-and-white, stilt-legged wader has an upturned tip to its bill. It leans down so that the tip is approximately horizontal and sweeps it to and fro through the water while making tiny dabbling (rapid up and down or side to side) movements. The technique sifts and traps small animals such as shrimps and worms. The flamingo has a similar but "opposite" method, since its bill is downturned. It holds its head to look almost vertically downwards, so that the main part of the beak again lies horizontally in the water. With a dabbling motion, comb-like structures inside the beak sieve tiny animals from mud and silt.

PAWS AND CLAWS

Some of the largest animal claws belong not to the big cats, but to the bears. A big grizzly's paws are almost the size of dinner plates, and the five curved claws on each foot are much larger than your own fingers.

Like dogs, and unlike cats, bears cannot pull in or retract their claws into fleshy sheaths at the ends of the toes. Unprotected from general wear and tear, the claws are not especially sharp. However, their size, and the immense power of the bear's limbs, more than make up for this drawback. Also, bears are plantigrade, which means that they walk on the soles of their feet, like us, and not on their toes. This saves the claws from too much wear and chipping.

Going fishing

Bears use their claws to grip when climbing, or when on slippery rocks or ice. They also dig for roots and soil animals with them, and slash at larger prey such as deer.

One of the most fascinating paw-and-claw uses is when fishing. In autumn, salmon produce their eggs (spawn) in the rivers, and grizzlies gather to feast on this highly nutritious food source. The bear stands on the bank or in the shallows and simply scoops up the fish in its forepaw, curling the claws around it to form a bowl-like fishing basket. Another technique is simply to jab and spear the fish with the claws, and then pick it up.

This fishing technique often looks more impressive than it really is. The salmon are exhausted after their long migration upstream and their spawning efforts. They are often close to death, stranded in the shallows, and are therefore a relatively easy catch for the bear.

LEFT Bears use many methods to catch fish, from hooking, slapping or crushing with a paw, to simply grabbing a fish in the shallows with the teeth.

HOW PREY PROTECT THEMSELVES

ABOVE Pangolins (scaly anteaters) live in Africa and southern and south east Asia. They lack teeth, but feed like the true anteaters, by licking up ants and termites with a very long, sticky tongue. Like armadillos, these mammals are well protected by their hard outer plates of keratin.

Predators have a huge range of weaponry to catch and kill their victims. But prey are far from helpless. Some have a similarly wide range of defensive armour to combat teeth, claws and beaks. A predator seeking soft, accessible flesh that is full of nutrients and easy to eat, will usually leave well alone when faced with a heavily protected creature.

- A hard outer exterior is one common form of protection. It is found in almost all animal groups, from mammals such as armadillos and pangolin, to reptiles like tortoises and turtles, to beetles and snails in the garden, and shellfish in the sea. Animals with this type of armour tend to be slow and heavy and may not be able to run away. They rely on a tough defensive armour to deter hunters.

- Spines and spikes are also successful anti-predator devices. They are found on porcupines, hedgehogs, porcupine fish and many others.

- Chemical defence is another widely used technique. The bombardier beetle can squirt a mix of stinging chemicals from its abdomen, which puts off even a hungry and determined bird.

- Chemicals coupled with spikes or spines can be doubly deterring. Lionfish have sharp fin spines that are extremely venomous, and some types of sea urchins are prickly, poisonous balls.

- Prey senses and speed, and warning coloration, all of which help prey to avoid capture, are described in other chapters.

LEFT The bombardier beetle is hardly larger than a fingernail, but it can spray a puff of harmful chemicals from its abdomen to deter enemies.

BELOW The chameleon's tongue, almost as long as its body, shoots out with unerring accuracy to trap a cricket on its glue-like tip. Its prehensile tail is wrapped around the twig for extra security.

TRAPPED BY A TONGUE

Apart from teeth, claws and beaks, many other body parts are used as offensive predatory weapons. Perhaps one of the least expected is the tongue. Certain types of tall-bodied, thin, slow-moving, highly camouflaged lizards known as chameleons have tongues longer than their bodies.

Normally the chameleon's tongue is folded and retracted into the back of the mouth. A hungry chameleon creeps unnoticed towards prey such as a fly, or sits still and waits for one to land nearby. Moving its turret-like eyes independently of each other, it assesses direction and distance. Suddenly the tongue shoots out, faster than our eyes can see. The fly has no time to escape as the tongue's saliva-sticky tip grabs it. The tongue is pulled with a sudden flick back into the mouth, and the jaws crush the catch.

THE JACKKNIFE KILLER

Forelimbs can also be deadly weapons, especially among smaller hunters such as insects and crustaceans. One of the former is the praying mantis. It may seem to have its front pair of legs folded as if in prayer, but they are designed to kill in a split second. Two long portions of each leg, the femur and tibia, lie hinged against each other, and each has spines that interlock with the other. The result is a combination of shears and sharp-toothed comb that snap together like a penknife blade flicking back into its handle.

Mantis vary in length from about one to more than 10 centimetres. Most are tropical, and are ambush specialists. They wait and watch, many disguised with astonishing realism as leaves, flowers or bits of twig. The prey draws close and – snap. In less than one-tenth of a second the front legs extend forwards, their portions straighten out and then immediately fold or close upon the victim, impaling it on the vicious spines. The prey is then pulled close to the mantis' mouth, where the powerful chewing mandibles get to work and tear its body into bite-sized chunks.

LEFT The mantis prefers to eat its victim – here a small cricket – beginning at the head end. This means the eyes and brain are consumed first, and so the prey ceases to struggle and kick.

A KNOCKOUT BLOW

The crustacean (crab and shrimp) group also has its fatal forelimb experts. Mantis shrimps or squillas are named after the praying mantis. There are some 200 kinds, varying from about two to 30 centimetres in length. At first glance they vaguely resemble lobsters. But their other common names of prawn-killers and split-thumbs describe their killing technique. The first pair of limbs on the elongated body are actually long, slender claws, in the style of a typical shrimp or prawn. But the next pair are large and powerful, with the jackknife design of the insect mantis. In some mantis shrimps the two portions of the limb have spines and jab into the victim, then snap together. In other cases one portion is straight and sharp-edged, and slots into a groove in the other. Guillotine shrimps would also be an apt name.

Like its insect equivalent, the mantis shrimp lurks, waiting in ambush – in this case, in a hole or rocky crevice in shallow water. Its strike is one of the fastest movements in the entire animal kingdom. A passing fish, prawn or shrimp can be impaled or even sliced in two in less than one-hundredth of a second.

Even more violent, and just as fatal, is the pistol shrimp or club-clawed shrimp. It has a pair of forelimbs that work less as pincers and more as clubs, with tips shaped like guns – the long part or barrel widens into the club-shaped handle at the tip. The pistol shrimp simply jerks its claw through the water at prey, and can smash the hard shell of a passing crab or similar victim with a single well-aimed blow. Captive pistol shrimps which engage in territorial battles with their own kind have been known to use their clubbed pincers to attack their reflections in glass – and completely smash the side of their aquarium.

BELOW In the warm waters of south east Asia, a mantis shrimp or squilla prowls through the shallows. Behind the long antennae, the second main pair of limbs are the jackknife weapons for grabbing prey.

WHOLE BODY WEAPONRY

A few predators use themselves as killing tools. The whole body becomes the instrument of death. Biggest are the constricting or squeezing snakes, the pythons and boas. The boas are found mainly in the New World (the Americas) and include the world's bulkiest snake, the nine-metre green anaconda, which weighs in at 220-plus kilograms. Pythons live in tropical parts of the Old World (Europe, Africa, Asia and Australia); at ten metres the reticulated python is the world's longest snake.

Some constrictors use their long teeth to grab prey. In fact the emerald tree boa has the longest, strongest fangs of almost any snake. But most constrictors use the more sinister technique of squashing their victim to cause death. They take advantage of their sheer weight, bulk and muscle power.

Lying in ambush, the snake strikes fast and bites the prey, such as a small deer or bird, to disable or distract it temporarily. Even as the victim starts to recover, the snake begins to coil around the prey's body. This is a strangely slow process. The constrictor is usually far heavier than its future meal and uses its weight to prevent the victim's escape. The coils snake around the prey and tighten.

The victim begins to gasp for breath. But each time it exhales, ready to take in lungfuls of fresh life-giving air, the coils increase their grip. The victim's chest cannot expand again, and the air never comes in. With each breath out, the body is squeezed smaller. The victim's heart is also affected very quickly. Under the immense pressure, it can no longer beat or send blood pulsing through the main vessels. Often, the victim shows little physical damage to suggest crushing, such as a stove-in skull or a caved-in chest. But slowly the breath is driven from the body, the blood is stilled, and life ceases.

LEFT An emerald tree snake slowly squeezes the life from an opossum. Growing to 120 centimetres in length, this South American rainforest snake can also strike with its strong fangs, either to kill the prey or to wound it before constricting.

RIGHT The thrush hops or flutters low in parks, gardens, woods and meadows, and picks up snails in its beak.

BODY OR BEHAVIOUR?

Evolution is generally a long, slow, gradual process. Hunters and hunted are involved in a never-ending "arms race". As predators try to catch, and prey attempt to escape, new body parts or behavioural techniques evolve. One example involves the snail, fairly well protected in its sturdy spiral-coiled shell. But where there is a meal, predators will, over time, evolve a means of overcoming the problem. For example, two types of bird have evolved different ways round this hard-cased defence.

■ The Everglades or snail kite of the tropical Americas and Caribbean region has an exceptionally long, slim, curved upper beak, which can poke through the shell's mouth (opening) and penetrate inside, to pull, lever and "winkle" out the snail. More often, however, the kite uses an easier option. It swoops down and uses one foot to pick up a freshwater snail. This is taken back to one of the kite's regular feeding stations. Still holding the snail in one foot, the kite waits for its head to emerge, and then suddenly slashes with its sharp beak. The snail is disabled and falls limp, and the bird can simply shake it free from the shell.

■ The thrush is also a keen snail-eater. But it does not have a specialized body part to winkle the flesh from the shell. Rather, it has a specialized behaviour. The thrush picks up the shell in its beak and smashes it against a hard stone. After a blow or two the shell cracks and the bird can peck through it for the slimy, but meaty body, inside. Empty, broken shells litter the area around the thrush's favoured smashing stone, which is known as its anvil.

A CURIOUS PUZZLE

A chainsaw might seem like a fine weapon, if slightly noisy and unwieldy, for cutting up the prey's body for eating. The animal version of this fearsome tool is the saw-like snout of the sawfish or smalltooth, a member of the ray and skate group of fish, closely related to sharks. The greater sawfish reaches an impressive six-plus metres in total length. Its snout may be one metre long, flattened with parallel sides, like a round-tipped blade. Its edges are studded with about 32 sharp teeth along each side. The teeth do not move and rotate around the edges, but apart from this detail, the resemblance to a chainsaw is quite remarkable.

But does the sawfish use its fearsome snout as a hunting tool? Perhaps, but not always, and not always obviously. Observing these huge animals in the murky shallows of the Atlantic coasts is difficult. There are several main theories about what the saw is for.

First, the saw may be mainly an offensive weapon. The sawfish slashes through shoals of small fish, wounding and dismembering them. It then returns to eat the injured and dead bodies.

Second, it is thought that the saw may be primarily a defensive weapon. It can be waved in display or used in anger to fend off enemies such as sharks, making the sawfish less vulnerable to attack.

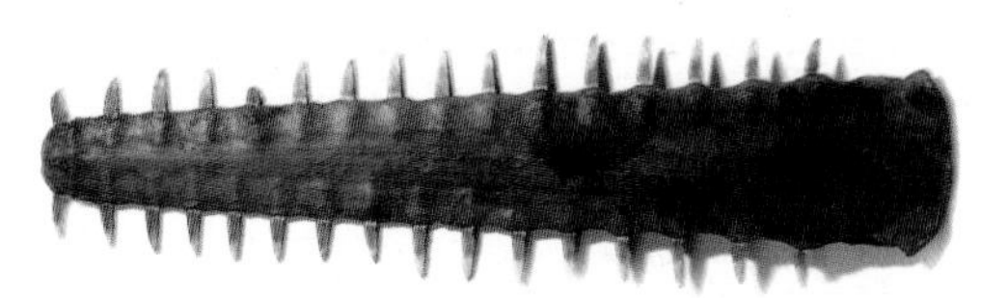

ABOVE The sawfish's snout teeth have same origins and structure as the mouth teeth. They are enlarged versions of the scale-like denticles which cover the skin of sharks and rays.

Third, and less impressively, the saw may be a combination of probe and lever. Sawfish seem to eat mainly small bottom-dwelling creatures such as worms, shellfish and flatfish. They have been observed to root up small prey from the mud or rocks of the ocean floor, using the tip of the snout.

Fourth, the saw may no longer have an obvious function. Long ago, it was important for some reason, to the distant ancestors of the sawfish. But conditions and environments change, and the modern sawfish now has no major use for its amazing nose. The saw is a "leftover of evolution".

The reality could be somewhere between all of these. The sawfish's chainsaw appears, to us, a perfect predatory device for large-scale slaughter. So why should it be used simply to poke worms from mud? No reason. We must be careful of transferring thoughts and ideas from our own daily experiences, dominated by machines and tools and gadgets, on to animals that live very different lives in very different places to ourselves.

BELOW Streamlined like a shark, but also flattened like a ray, the sawfish feeds on or near the bottom.

ABOVE The sea otter's tool is one of the most common objects in its environment – a pebble from the sea bed. The otter uses the pebble to crack the hard outer casing of the shellfish to expose the meat within.

MAKE YOUR OWN WEAPONS

Some predators do not have hunting weapons as parts of their bodies. Instead, they adapt items from the environment. They are tool-users.

- The sea otter of Pacific coasts dives to the sea bed for prey such as fish, crabs and shellfish. If the prey item has a hard body case, like an oyster or clam, the otter also brings up a pebble from the sea bed. It lies floating on its back, puts the shellfish on its chest and hits it repeatedly to smash open the shell.

- The Egyptian vulture also uses a rock as a hunting tool. The vulture picks up a stone in its beak and passively drops or actively throws this at a bird's egg. After a few attempts the stone cracks the shell and the bird can peck up the egg inside.

- The archer fish's weapon is all around – water. It spits or squirts an "arrow" of water from just under the water's surface, to knock small creatures from overhanging leaves and twigs into the water below (see back cover image).

5 TEAM SLAYERS

In the northern wilderness, the howl of a grey wolf sends a chill down the spine. If a person came face to face with this powerful predator, what would be the outcome? Very probably, the wolf would retreat. Despite centuries of myth and legend, a lone wolf is extremely unlikely to attack a human being.

The howls of a grey wolf pack send even more chills down the spine. Compared to a solitary wolf, the pack is a different matter. Again, despite stories of folklore, they are very unlikely to approach or attack a human, unless perhaps ravenous with hunger. But the wolf pack is a prime example of killing by cooperation. Wolves in a pack work together to bring down prey as big as a moose, which could feed the whole group for a week.

STAYING FLEXIBLE

Why do some predators operate as team slayers rather than lone hunters? This is a complicated question and there is a whole range of examples, from animals that never hunt with others of their kind, to a few which almost always do. However, in most cases, the predators show very flexible behaviour. If a small prey item becomes available, an individual snaps it up. If a large quarry is detected, the group band together to pursue it.

Some predators plan group attacks. Lions, wolves and killer whales hunt in an intelligent way, even predicting prey movements and preparing to cut off their escape routes. As they cooperate and coordinate their attack, communicating with signals such as movements and sounds, they show division of labour as individuals take different roles for the hunt and kill. In this way, they can catch bigger prey or more victims than each could manage alone.

But hunting in a team has its disadvantages. Because the kill is usually shared out, each predator may end up receiving less than it would gain from sole effort. So when is it worth teaming up, and when is it better to hunt alone?

One theory as to why predators hunt in packs is known as Selfish Gene Theory and concerns relatedness and genetics. If members of a predatory pack are closely related, such as siblings (brothers and sisters), or parents and offspring, then they have many genes in common. Working together for the good of the group, and doing more than your fair share, may help your close kin to survive, even if you do not. This increases the chances that your genes will be carried to the next generation. The genes are passed on by your kin.

OPPOSITE Grey wolves pick over a deer carcass. In winter, when food is scarce, the bones are crunched and the hide chewed.

BELOW Lions rest by a carcass that vultures and other scavengers have already picked at. These big cats are not above scavenging themselves.

SIDE BY SIDE, BUT NOT TOGETHER

Wildebeest trek through the African bush on their seasonal migration, to find regions where recent rains have encouraged grass growth and fresh grazing. They come to a river, pause warily, but then plunge in, wading and swimming hastily to the safety of the other side. In the water lurk Nile crocodiles – dozens of them.

Are the crocodiles team hunters? Not really. This is an example of a situation in which predators group together because prey are also gathered. The hunters take advantage of the aggregation of easy meat. There may be some elements of cooperation. Crocodiles rarely battle over exactly the same piece of flesh on a carcass. They usually bite at different sites or take turns to pull and tear chunks off a body. But in the actual hunt, it is a solo predatory effort, with every crocodile for itself.

ABOVE Crocodiles often congregate in large numbers, drawn to one place by large concentrations of prey, or to bask in the sun on a river sandbar or in a muddy pool.

SWAMPED

There are many other instances of carnivores which group together where there are large numbers of their prey. Many kinds of dolphins form large schools of hundreds of individuals, as they attack vast shoals of fish or squid. The individuals in the school communicate and socialize in various ways, especially for breeding, navigation and "play". But on the hunt itself, apart from a few notable exceptions (see below), they seem to operate largely alone. Many predatory fish that themselves live in shoals, such as marlin or tuna, attack shoals of prey like herring or mackerel. Groups of penguins feed in a similar way, dashing about through the water with their paddle-shaped wings after densely packed food such as small fish and shrimp-like krill.

Crocodiles, dolphins, penguins and others take advantage of the so-called swamping effect. Since there are many predators, the prey become confused and panic. Chaos reigns. A prey may swerve or sprint to avoid one hunter – straight into the mouth of another. For a short time, there seems no escape as the feeding frenzy is in full swing.

BELOW Penguins dash around underwater causing their prey, usually a shoal of fish, to panic. This type of cooperative hunting enables them to catch more prey than when hunting alone.

THE FEEDING FRENZY

Piranhas indulge in feeding frenzies, but of a different kind. The red-bellied piranhas of South America are about 30 centimetres long and inhabit lakes and slow rivers. They have extremely strong, deep jaws, armed with rows of thin, triangular, blade-like teeth. A piranha on its own would probably not refuse a small creature such as a baby bird, frog or young fish. But in a group they are known to tackle animals as big as deer, tapirs, horses and occasionally even humans.

However, piranhas are less an example of coordinated pack hunters, and more a shoaling collection of individual predators. They rarely scavenge, but eat only fresh flesh. They also rarely attack healthy animals that move purposefully through the water with regular motions. Instead, they are attracted by the struggles and panic splashes of an animal in trouble – wounded, sick or drowning.

Very cautiously, the piranhas gather round a potential victim. A few smaller, younger members of the shoal begin to dart in from the sides and take quick test bites. These younger fish are very fast and agile and, compared to the adults, make a smaller target if the victim is able to fight or bite back. If the trial attacks are successful, then larger, older piranhas join in. Each bite slices out a lump of flesh approaching the size of a human thumb. There is little detailed coordination or cooperation between members of the piranha shoal, but their massed efforts are irresistible and extremely successful. As the frenzy mounts, the fish begin to bite almost anything within reach. The victim is stripped of all flesh within minutes, to leave the clean-picked bones of the skeleton.

BELOW Piranhas lurk in a shoal just below the surface. They look, smell, and also feel the ripples and currents in the water. If one detects a possible victim and moves in a certain direction, the others will follow.

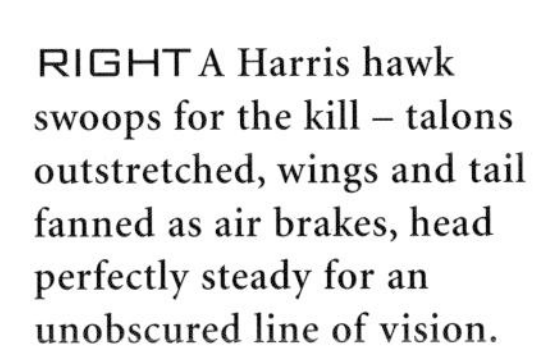

RIGHT A Harris hawk swoops for the kill – talons outstretched, wings and tail fanned as air brakes, head perfectly steady for an unobscured line of vision.

A PACK OF BIRDS?

There are very few examples among the birds of true pack-hunting behaviour. One is the Harris hawk, a medium-sized raptor from the deserts of America. As light dawns, a small group of these hawks, usually three or perhaps four, gathers on nearby perches to survey the countryside. They take turns to soar around and spot prey. Once a victim is located, such as a rabbit or rat, the group can adopt one of several options. They may take turns to swoop down and harass the prey, until it becomes tired or confused and easier to catch. They may indulge in a mass attack, gliding in from all sides, talons at the ready. Or one of the hawks may land and part run, part hop and part fly along the ground, for a surprise ambush. When the prey is secured, the birds take turns to tear off lumps of meat. Harris hawks also show unusual breeding behaviour among birds, known as polyandry, where one female courts and mates with several males.

PRIDE IN PREDATION

The mammal group contains the most complex and exciting examples of true pack-hunting predators. Most of these are canids – members of the wolf and dog group, as described below. Chimpanzees also show cooperative hunting behaviour in their raiding parties (see page 120). But perhaps the most extraordinary case comes from a group in which almost all other members are solitary stalkers – the cats.

A "pack" of lions is known as a pride. It usually consists of between four and six lionesses (females) and their cubs, plus one or two males. The lionesses tend to be close relatives such as mothers, daughters, aunts and sisters. The males may be related to each other, but not usually to the females.

The male lion is massive and impressive, with his long head-and-shoulders mane and fearsome roar. However, he hardly ever hunts. His main tasks are to patrol the pride's territory, to keep out rival prides and other enemies, and to mate with the females.

The lionesses do the hunting. They usually set out as a group as darkness falls. They tend to have different roles, often according to their size and build. The lighter, faster lionesses are the pursuit specialists – "flankers" or "wingers". They locate and stalk prey, then charge towards it at sprint speeds of more than 55 kilometres per hour. The stronger, heavier lionesses have the job of springing the ambush. They are the team's "centres" or "catchers", who leap from cover to grab and bring down the fleeing victim. As they grapple with the kicking hooves and jabbing horns, often hanging on to the muzzle or neck, the other lionesses rush up and begin the disembowelling process.

This is the story of an idealized pride hunt. Working together, lionesses can bring down large, powerful prey such as adult wildebeest, zebra and even the huge and aggressive African buffalo, with horns longer than your arms. But the lion is flexible and opportunistic, with a wide diet and range of prey. The roles in the hunt may change if the prey makes an unexpected dash or if a new quarry appears. Or an individual lion may just as easily hunt alone for small prey such as rats, hares, snakes, frogs and lizards.

BELOW Only lionesses are tackling this African buffalo – one of their most powerful opponents. The male lions will stroll up once the kill is secured and take the prime meat.

THE CLASSIC PACK

The grey wolf (timber or "common" wolf) is an impressive beast. A large male measures almost two metres from nose to tail, is 70 centimetres tall at the shoulder and weighs 50-plus kilograms. Several of these wolves together are even more impressive, indeed extremely intimidating, as they snarl and bare their long canine teeth. Once a victim is cornered by such predators, there is little chance of escape.

As in most larger members of the dog family, the wolf pack is a loose, extended family group. It may number up to 30, although eight to 12 (including young) is more usual, where prey is plentiful and human persecution is minimal. In smaller forest areas there may just be one breeding pair and their cubs per pack. In larger groups there is still only one breeding pair, the alpha male and alpha female. They are the true leaders of the pack, deciding when to hunt, when to rest and when to move on. Other adults in their pack are inhibited from breeding unless they successfully challenge the alphas and take their place.

Each pack has a territory, ranging from less than 100 to more than 1,000 square kilometres, again depending on factors such as the abundance of prey. The members of the pack patrol their territory regularly and threaten other wolves who try to invade.

ABOVE A grey wolf bares its teeth in a snarl, to keep a strange rival – in this case, probably the photographer – away from its meal. Food-stealers such as bears or pumas are a constant hazard in the harsh north.

Wolves on the prowl

The wolf pack's focal point is the rendezvous area. This is usually an open patch of land with good views all around, where the members gather to socialize, eat, rest and play with the cubs. The active adults collect here before setting out on the hunt, leaving suckling mothers or sibling baby-sitters behind. The hunting party may howl, to help bind the pack together and perhaps increase motivation for the coming chase. (Like a sports team's pre-match "pep talk" from the coach, some kind of pre-hunt motivational behaviour is common among canid packs.) Then the wolves sniff the ground and air for traces of prey. Once they are on the scent of a target in this way, they rely on stamina more than speed. They can trot for many hours. A pursuit of more than 30 kilometres is not exceptional.

BELOW Usually the wolf pack's leading pair, the "alpha" male and female, have first choice of food. In times of famine, pack members may threaten and fight each other. Food then becomes a basis for challenging pack dominance.

On the other hand, prey may be much nearer, such as a herd of caribou (reindeer) or red deer. In these cases the wolves quickly assess the fitness of the various individuals. They do not waste time and effort chasing those which are strong and healthy. More likely they single out an easier target such as an old or sick individual, or a youngster temporarily separated from its mother. This can be done by a series of quick test attacks. Possible prey that run away quickly are ignored.

Closing in

As the pack closes in on the victim, there is again a choice of strategies, depending on the circumstances.

- The younger, lighter, faster but less experienced wolves may drive the chosen prey towards the bigger adults. One of these larger wolves might leap or lunge, and try to trip or knock the victim, so that other pack members can come in rapidly and help to subdue it.
- Another aim, especially with larger prey, is for the pack to form a circle around the animal. Then one of the alpha pair may dart in and grab its nose or back leg. Again, this is the signal for the massed attack.
- If the prey tries to break out of the cordon and escape, the nearest wolves dash at it and snap at its flanks and legs.
- A relatively slow but powerful adversary such as a bison may be chased in dogged pursuit for an hour or more. The wolves attack at intervals, snapping and ripping, causing wounds and tearing out small lumps of flesh. Finally the animal collapses from exhaustion and loss of blood.

Whatever the details, once the prey is caught, it is usually dead in a few seconds.

As a pack, wolves catch prey as small as the young of deer, mountain sheep and mountain goats, up to adult caribou, elk, moose and bison perhaps three times the pack's combined body weight. As a lone hunter, the wolf is also exceptionally adaptable, with many other approaches. But it is less bold and more cautious than when it is in a group. It tends to crouch low, creep along the ground, get as close as possible, and then dash towards or spring at the prey. Hunting alone, it usually tackles only small prey such as mice, voles, rats, ground squirrels, rabbits and hares. Wolves also cache – hide or bury excess food for later. A pet dog which buries its bone in the garden has inherited the behaviour from its long-distant wolf ancestor.

MORE PACK HUNTING DOGS

The coyote has a reputation as a lone predator, fostered by its mournful solo howls in tales of the American Wild West. It resembles a wolf and is a very close relative, but it is smaller and lighter in build, perhaps 130–140 centimetres in total length and 15–20 kilograms in weight. Coyotes have expanded their range in North America, through Mexico to northern Central America, partly due to their habit of hunting farm animals such as sheep and chickens. They have also adapted well (often too well) to human habitation. They raid gardens, storerooms and garbage for almost anything edible, even fruit, berries and kitchen scraps.

Recent studies show that coyotes usually form breeding pairs or live alone, but in some cases, a pack forms. This tends to be a family group where the young have reached adulthood, but have not left or dispersed in the traditional way. The original parents become the alpha female and male, and are the only ones to breed.

Alone, a coyote is very cautious and preys on small items such as rats, mice, ground squirrels, lizards, beetles and other insects, bird eggs and chicks. But a pack becomes bolder and pursues deer, pronghorn and mountain sheep. In some cases they intimidate people and even consume domestic pets such as dogs and cats. Reported serious attacks on humans, especially in western North America, are now running at more than five per year.

The dingo or Australian wild dog has a similar chequered history. Some experts believe dingoes were once fairly well domesticated, but they have gone semi-wild or feral and now roam at large. This is

THE RED WOLF

Many close relatives of the grey wolf show similar group behaviour and pack-hunting techniques. The red wolf is slightly smaller and slimmer than the grey wolf, and its range was mainly south-east North America. But it probably became extinct in the wild in the 1970s, due to habitat loss, persecution by humans and other causes. Eight captive-bred individuals were reintroduced into North Carolina in 1988, and a few other selected introductions have occurred since. As in the grey wolf, the red wolf pack consists largely of related family members. They can hunt together to bring down prey such as deer or wild pigs, but they also tend to forage alone for rabbits, racoons, muskrats, ground birds and similar smaller items.

LEFT Dingoes are members of a group of dogs found from North Africa and Eastern Europe, right across Asia and down into Australia. They live alone or in packs, and are wild or semi-domesticated.

ABOVE A small coyote pack surveys a flock of snow geese, looking for weak or injured birds that would be easier targets.

especially true in the islands of Southeast Asia, where dingo packs hang around towns and villages, and may interbreed with domestic dogs. Flexible and opportune in its diet, like the wolf and coyote, the dingo likewise becomes bolder and more of a menace if it is part of a pack, which may be 10 in number. The pack hunts large prey such as wallabies and kangaroos, and there are regular reports of people being menaced, with the occasional fatality.

The dhole or Asian wild dog is another highly social canid. It lives in extended family packs of up to 25, although five to 12 is more typical. Working together, dholes can catch chital and sambar deer far larger than themselves. They tend to hunt in packs more often than alone.

The bush dog of South America is less like a typical wolf, and more like a combination of dog and stoat, long and bendy. It measures only 80 centimetres from nose to tail. Individual bush dogs prey on rodents such as rats, ground squirrels and agoutis (cousins of guinea pigs). But a pack of 10 or more will challenge the tall flightless bird called the rhea, which is South America's version of the ostrich, or even the capybara – the world's largest rodent, resembling a long-legged guinea pig, and almost the size of a small pony.

OPPOSITE Asian wild dogs, or dholes, tend to hunt in packs more than alone. They also make communal dens with more than 30 metres of tunnels.

BELOW African wild or hunting dogs are hardly ever found alone. They are the most social of the canids (dog and fox family). They actively support fellow pack members who are unable to fend for themselves.

A CUT ABOVE THE REST

The prize for the most efficient pack hunters goes to the African wild or hunting dog, which can bring down the largest prey compared to its own body size. An account of a typical hunt is shown on page 93.

The African wild dog is about 120 centimetres from head to tail and weighs 20–30 kilograms. (Its prey may exceed 1,000 kilograms.) It is lean, rangy and speedy, with very long legs – the dog version of the cheetah. This species is extremely threatened, with perhaps only 5,000 left in scattered regions of Africa including Tanzania, Zimbabwe and Botswana. It has been relentlessly persecuted, shot, trapped and poisoned, is the victim of traffic accidents and deliberate road kills, and has been affected by diseases such as canine distemper and rabies caught from domestic dogs.

The African wild dog pack ranges in size from five to fifty, but a typical group has about five to nine adults and six or seven pups. As in wolves, only the dominant or alpha pair breed. These dogs occasionally hunt for small prey such as rats and rabbits. But more often they opt for mass pursuit of larger game, including gazelles and antelopes such as springbok, impala and kob, and even wildebeest and zebra. The pack runs down its quarry at high speed, and shares the food among the pups. Even injured pack members, who cannot keep up with the main chase, and would not survive alone, are encouraged to follow and feed. In some regions, African wild dogs are successful in four out of five pursuits – an amazing success rate.

WHY PREY STICK TOGETHER

Predators such as wolves and wild dogs usually have greater success when they hunt together. But why do prey stick together too, in such large and easily spotted herds, flocks or shoals? The answer is that they have safety-in-numbers behavioural strategies to avoid being eaten.

- In a large group, there is more likelihood of early predator detection. While some prey eat or rest, others can be on the lookout, listening keenly or scenting the air for danger.

- If the enemy is detected, and other members are nearby, a warning sound or movement can alert the whole group. Such warnings would travel less effectively between individuals spread out over a large area.

- In a big group, an individual prey is less likely to be singled out and killed. In particular, small fish in a shoal, or birds in a flock, seem to move and wheel in perfect unison, as one huge "super-organism". A predator may become confused and unable to focus on one individual to attack.

- In some prey groups, members may band together to repel enemies. They work as an anti-pack to chase off hunters.

THE BIGGEST PACK – THE POD

Wolves or even lions pale into the background when faced with the biggest pack hunters on Earth – killer whales or orcas. These massive predators dwell in all saltwater habitats, from estuaries and shallow bays to open oceans, especially in cool to cold waters, and polar seas with icebergs. A large male orca is almost 10 metres long and 10 tonnes in weight, with a dorsal fin as tall as an adult human; the female can be more than half this size. The 10–12 pairs of peg-like teeth in both the upper and lower jaw are adapted to seize any prey, and escape is unlikely since the orca surges through the water at a tremendous rate, approaching 50 kilometres per hour – faster than most other marine animals. Each individual has a unique pattern in its black-and-white coloration.

Killer whales are the largest members of the dolphin family and live in close-knit, long-lasting groups called pods. A typical pod seems to be matriarchal, that is, based around a senior female. Both her male and female offspring tend to stay with her, and so on, so that several successive generations build up around the original mother. However, males leave occasionally to mate with females from other pods, thereby preventing inbreeding.

Some pods seem to stay put in an area. These "residents" specialize in eating fish such as salmon, also penguins and other sea birds. Other pods, the "transients", are usually on the move. They tend to prey on marine mammals such as seals, porpoises, dolphins and whales.

ABOVE A killer whale (orca) dwarfs a roughly human-sized member of the same family, a dolphin. Even solitary killer whales are capable of taking on almost any prey in the ocean, apart from the great whales.

Many ways to kill

The killer whale has a huge range of hunting strategies, which it carries out in a flexible fashion, alone or with others in its pod. The pod may herd shoals of fish or squid upward, trapping them against the surface of the sea, which is a roof for them but not for their air-breathing pursuers. Then the killers swim through and head-butt or snap up the victims, or stun them with powerful flicks of their muscular flukes (tail fins). Or the killers may breach – swim quickly upwards, leave the water and crash down among a shoal near the surface, stunning their victims.

Two or three killer whales may chase smaller dolphins, porpoises, seals or penguins into a steep-sided bay which acts as a dead end. If the seals or penguins leave the water and take refuge on the rocks, the killers flick great quantities of water at them with their flukes, and try to wash them down and back into the sea. Should seals or penguins seek safety on an ice floe, the killers surge up from below and smash the ice or tip over the berg. Another trick is when the killer surges up on to the beach, as though surfing, and grabs a seal or penguin before turning round and wriggling back into the water. A pod may even chase a great whale such as fin, sei or blue, grabbing mouthfuls of flesh as they pursue.

BELOW Killer whale dorsal fins break the calm surface as the pod cruises unhurriedly towards its regular feeding ground. But they will divert from their course to take advantage of a chance meal.

One scientific study observed killer whales on more than 550 hunting events. More than 200 of these revealed two or more killers herding, corralling or otherwise catching prey together. They seem to communicate using body posture such as an arched back or flippers held in a particular position and by swimming styles and speeds. They are also thought to share information vis a vast range of sounds including clicks, screams and whistle pulses. These noises increase in number during the hunt, and some are used for echolocation (see page 28). Although we have not yet been able to translate killer whale sounds into human language it is known that killer whales from different regions have their own distinct dialects. Humans still have a huge amount to learn about these champion pack hunters of the vast oceans.

6 CHEATS AND PART-TIMERS

Some predators do not hunt their own prey. They cheat by stealing it from another. They attack, confront or challenge a predator which has already secured a victim, with the aim of taking the food outright, or sneaking it away using some kind of trickery. These animals are the burglars, cheats, pirates and even parasites of the predatory world. They include various kinds of spiders and, in particular, several types of birds which are both robbers and actual predators in their own right.

MUGGERS AND REAL PIRATES

The pirate spiders, or mimetids, may sound as if they raid the webs of others to steal their prey. But this particular pirate is misnamed. It is a bulky and slow spider, but exceedingly stealthy and cautious. It does indeed invade the web of another spider, but not to carry away its victims. Instead the pirate spider creeps towards the web owner, very slowly and smoothly to avoid alerting it – and then kills and eats it.

The kill is usually made when the pirate has readied itself in a suitable place on the web. It pulls on one of the silken lines in the web, to attract the owner. The web-making spider approaches in the usual manner, tricked into thinking that its sticky net has trapped a small victim such as a fly. But the pirate surges forward in a surprise attack and bites the leg of the web owner hard. Its fangs inject powerful venom that works so fast that the resident spider is dead within a second or two. Such fast-acting poison avoids a struggle between the two spiders, which would be very risky for the pirate.

Some types of crab spider may take a chance and creep around the web of another spider, looking for uneaten prey trapped there. The crab spider usually does this when the web owner is busy tying up or feeding on another, larger meal in a different part of the web. Again, the crab spider moves very slowly and cautiously, to prevent detection by the web owner, and also to avoid getting caught itself.

OPPOSITE An Arctic skua harasses a kittiwake with pecks, wing-flaps and kicks – forcing it to release its hard-won food.

BELOW Crab spiders usually lurk camouflaged in flowers, waiting to seize nectar-feeders such as small flies and beetles. But they also creep around the webs of other spiders to see if there are any easy pickings.

SMALL-TIME THIEVES

The expert prey stealers in the spider group are known as commensal spiders. The term "commensal" refers to interactions between individuals from different species, where one gains a benefit or advantage, and the other is not harmed or disadvantaged. (Compare commensalism to parasitism, see page 17.)

The small commensal called the argyrodes spider usually lives in the large orb web of a much bigger spider. The size difference is so marked that the web owner is not really concerned about the little intruder, which is hardly worth throwing out or hunting as food. Like a human squatter living in a building but paying no rent, the argyrodes spider lurks around the edges of its host's web, staying out of the way and trying not to cause any trouble. Its food is prey trapped in the web – but prey which, again, is too small to bother the host. In this way the two spiders coexist, one gaining seemingly at no expense to the other. However, it has been suggested that the web-owning spider may benefit because the commensal keeps its web tidy and free from small bits of clutter and debris. If this is the case, the relationship is not so much commensalism as mutualism or symbiosis, where both partners gain an advantage.

Another commensal is the uloborus spider from tropical Asia. It feeds on flies and similar prey trapped in the large web of a social type of spider known as the stegodyphus spider. Several of these social spiders build and maintain the web, and the uloborus spider keeps a low profile on it. In fact the commensal not only feeds in the social spider's web, but carries out its whole life cycle there. The female uloborus lays her egg sacs in the silk strands, and the baby uloborus spiders hatch out and grow up in this hazardous home.

BELOW The fluffy webs of communal spiders may cover whole bushes or even trees. Members of the commune jointly catch larger prey and share it. But elsewhere in the huge trap, commensal spiders are picking off smaller, more incidental catches.

DAYLIGHT ROBBERY

Bird pirates are much more upfront than spiders about their stealing habits. Frigate birds soar for long periods of time across all tropical oceans. They are about one metre from the tip of the bill, along the light, slim body, to the end of the tail, and well over two metres from wingtip to wingtip across the narrow wings.

A frigate bird feeds for itself by gliding low over the waves and dipping its long, hook-ended beak into the surface. It grabs prey swimming or floating there, such as fish, squid, jellyfish or surface-dwelling shellfish. But the frigate bird is also an aerial hijacker. It swoops to and fro along the coast, watching out to sea for incoming birds such as boobies, gulls or petrels, returning with their recently caught food. Usually, this food has already been swallowed, but the frigate bird is not deterred. It worries and harries a victim by diving and flapping at it, and trying to peck its wings. In response the victim regurgitates or brings up some of its recent meal. The frigate bird dives and catches the disgorged food as it falls through the air, or swoops down and plucks it from the surface of the ocean as the messy meal splashes in.

Frigate birds rob not only birds. They also steal from large predatory fish such as marlin or tuna. As these big fish attack a shoal of smaller prey fish, the prey are bitten or wounded or stunned, and some float to the surface. Even as the predatory fish is about to swallow its rightful meal, the frigate bird swoops past and snatches the fish from its jaws.

ABOVE A magnificent male frigate bird inflates his throat pouch as part of his courtship display. Frigate birds both catch their own meals and steal those of other birds.

ABOVE Skuas endlessly look out for free food. The victim here is a king penguin chick, nicknamed an "oakum boy". It may have died naturally – or have been attacked by another predatory seabird.

More piratical birds

The skuas, also called jaegers, are another group of bird-harassing pirates. There are about five skua species and they resemble gulls with dark plumage. Largest is the great skua, at 60 centimetres from beak to tail. Like the other skuas, it sometimes hunts its own victims. It is powerful enough to kill ducks, gulls and similar coastal birds, and it is especially fond of bird eggs and chicks as it raids breeding colonies. It also takes carrion along the shore, such as washed-up fish, shellfish and injured sea birds. And it follows ships and consumes any leftovers thrown overboard.

The great skua excels at thieving. It is an aggressive bully and, like the frigate bird, it harasses other birds in flight. It tries to force them to drop or disgorge their prey, which the skua follows down and grabs in mid air with its long, hook-ended beak. The skua's techniques include flapping and pecking at the victim, or raking and jabbing with its sharp foot claws. One favoured method is to grab the end of a gannet's wing in its beak, so the gannet loses flight control and crash-lands in the sea. The victim regurgitates its food and the skua snaps it up. Other common skua victims include terns, gulls, kittiwakes and puffins.

Why do the victims of these bird pirates give up their hard-won food? Perhaps it is a defence mechanism when threatened – various kinds of birds throw up their smelly last meal if confronted by an enemy. Or the bird may be trying to lighten itself, so that it can escape the attack. Possibly it is done to provide the pirate with what it wants.

The arctic skua is another enthusiastic in-flight robber. Its scientific name is Stercorarius parasiticus, indicating that it parasitizes other birds by purloining their prey. One of its main quarries is the arctic tern. The arctic skua is also keen on the eggs and chicks of other birds at breeding time. Sometimes a skua pair work together on the attack. One distracts or challenges the parent, while the other plunders the nest. Another species, the pomarine skua, upholds the skua tradition by chasing terns or kittiwakes and making them release their food. However, it too catches its own fish, bird chicks, lemmings and other "legal"prey.

PART-TIME PREDATORS

Some predators exist on flesh alone. They must consume animal matter or they perish. Snakes, frogs, spiders and dragonflies follow this obligate carnivore or "must eat meat" lifestyle. But many other predators, especially among the mammals, are more adaptable. They eat flesh when it is available. Otherwise they scavenge or eat fruits and other plant matter, rather than starve.

Bears are members of the Carnivora mammal group (see page 63). Yet some bears, like the giant panda, hardly ever eat meat. The eight bear species represent the whole range of diets, from carnivore to herbivore. Three examples show how these huge, heavily built creatures arrange their menus according to what is available.

ABOVE Giant pandas have been viewed by experts as members of the bear family, or as a separate group within the Carnivora group. In either case, they hardly ever eat any meat. Their main diet is a type of grass – bamboo.

THREE OF A KIND

The polar bear is the most carnivorous of the bears. Its diet is almost exclusively meat, especially seals such as ringed, bearded and harp seals. It also catches the occasional walrus, sea bird and fish. It may scavenge on injured or stranded whales or porpoises. Some polar bears move inland and eat arctic hares, birds and their eggs, and even lemmings and voles. They sometimes pick over the remains of carcasses such as caribou or musk ox, usually left by wolves. In a few cases, polar bears have been seen chewing mosses, lichens, berries and leaves. But such signs of herbivory are rare, and are perhaps the desperate last measures of an old, sick or starving individual.

ABOVE The brown bear's long claws can snatch fish from streams or tear apart a deer. But the bear also uses them to scrabble in soil for fallen fruits and to dig up roots.

The brown (grizzly or Kodiak) bear is much more of an omnivore. It consumes fresh meat in the form of deer or wild sheep and goats. It has even been known to tackle a bison, yet also dabbles in a variety of small animals such as frogs, voles and ground squirrels. The brown bear will snatch fish like exhausted migrating salmon from streams, but it will also dig for roots and bulbs with its long claws, nibble tender shoots and buds and feast on berries, fruits and nuts in autumn. In coastal areas, it even chews on seaweed. When times are hard, the brown bear refuses hardly anything edible.

The spectacled bear is the only bear species in South America. It dwells in trees, and (apart from the giant panda) is the most herbivorous of the group. Its very strong facial muscles, sturdy jaws and large cheek teeth are suited to mashing tough vegetation. The spectacled bear picks part-time at small mammals, insects, birds, eggs and carrion. But its main diet includes fruits, berries, shoots and soft stems, as well as juicy leaves. It digs up bulbs and succulent roots, and pokes its nose into flower hearts to lick out the petals and nectar. It is even a farm pest, not so much for taking livestock, but for raiding crops, especially maize.

As with most herbivores, the spectacled bear's diet is plentiful, but relatively low in nutrients, and more difficult to digest compared to meat. It is estimated that the spectacled bear eats for between five and ten times longer, per day, than the polar bear.

Certain other mammalian carnivores show this type of adaptability. Foxes and especially badgers are flexible in what they consume. Unlike some humans, when the burgers and sausages run out, they readily switch to the peas and greens.

BELOW The spectacled bear is one of the most arboreal (tree-dwelling) of its family. It searches through the branches for the sweet, succulent centres or "hearts" of bromeliads and other flowering plants.

7 THE GAME OF DEATH

The black-footed ferret, from North America's Great Plains, feeds only on prey which it has freshly caught and killed itself. Indeed, it eats only one kind of prey, namely prairie dogs. In fact, black-footed ferrets are closely associated with prairie dogs in many ways. In the wild, the ferrets eat almost nothing else. A black-footed ferret measures about 50–60 centimetres from nose to tail, but is slim and bendy and weighs less than one kilogram. It is thin enough to use abandoned prairie dog tunnels for shelter, and as a bolt hole during its travels, to avoid danger. At breeding time the female ferret may enlarge an old prairie dog burrow as a den for her young.

The black-footed ferret is a very specialized predator. Provided prairie dog numbers remain high, the ferrets breed well and are successful. Predator and prey survive side by side: they have been coexisting for thousands, perhaps millions, of years on the great grasslands of North America.

NOTHING TO EAT

Since humans took over the plains for farming the number of prairie dogs has declined sharply. In some US states the prairie dog population has been slashed to less than one-fiftieth of its original size by trapping, poisoning, shooting and other means. This has helped the farmer, but not the black-footed ferret, which controls prairie dog numbers by more natural means. Unable to switch easily to other prey, and lacking its ready-made shelters and breeding burrows, the ferrets too suffered dramatic decline. This was worsened by a plague-like disease that wiped out yet more prairie dogs, and canine distemper, which somehow spread from domestic dogs into the wild ferret population. By 1990, black-footed ferrets were thought to be almost extinct in the wild, at least in the USA. Some may survive in Canada where they feed on ground squirrels.

Planning for such a disaster, conservation programmes had been started by breeding the ferrets in captivity. During the 1990s, several groups were released back into the wild at various sites in Wyoming, Montana, South Dakota and Arizona. They seem to be holding their own, but the future of the black-footed ferret hangs by a thread.

OPPOSITE Long and lithe, the black-footed ferret has little trouble invading prairie dog tunnels in pursuit of its prey.

BELOW Prairie dogs sit on a mound, looking out for predators both in the air (e.g. hawks) and on the ground. Their main defence is to flee into their tunnel system. A ferret can easily follow, but will take only a single prairie dog, so the rest of the colony survives.

SIMPLY A SCAVENGER?

The ecology of a predator describes its survival in the wild. It includes features such as how it copes in various habitats, the type of prey it eats, how it interacts with other predators, and its general relationships with the creatures and plants around it. Contrast the fairly simple ecology of the black-footed ferret – limited habitat, limited prey – with that of a much larger predator, the spotted hyena.

Hyenas may look broadly like dogs, from the outside. But they are more closely related to cats, civets and mongooses than to the canids (dog family). The hyena has a back that slopes down from its front to its rear limbs, a large head, massive teeth and extremely powerful jaw muscles. For its size, the hyena probably has the most powerful bite of any mammal predator. It can chew almost any part of a prey's body, including its hide (skin), gristle and even bones, which it cracks open for the soft, nutritious marrow within (see page 62). This is why the hyena is regarded mainly as a scavenger, hanging around while a big cat or crocodile eats the fill of its kill. Then the hyena slips in and chews over the tough remains.

The flexible approach

Largest of the three kinds, or species, of hyena is the spotted hyena. It is almost two metres from nose to tail and weighs 60–70 kilograms. Its main habitat is dry, open scrub, grassland and scattered woodland across much of Africa south of the Sahara Desert. Spotted hyenas tend to live in groups called clans. These groups vary in size from only four to five individuals in harsh almost desert habitats, to 80 or more in prey-rich habitats such as scattered bush and grassland. The clan has communal dens, marks out its territory and chases away invaders, especially other hyenas.

The spotted hyena is an extremely flexible predator. It can hunt alone, catching small rodents such as ground squirrels and rats, also hares, ground birds, and even fish and freshwater crabs. If it comes across the mostly eaten carcass of a large animal, it will chew and consume the parts which were too tough even for other scavengers, such as vultures and jackals.

Spotted hyenas also hunt in packs. They bring down antelopes, gazelles, zebra and similar large prey and occasionally cattle or other farm animals. Like other pack hunters (see page 88), the hyenas are very noisy as they prepare for the chase. They scream and howl to help encourage or motivate each other. In some regions of West Africa, spotted hyenas are more successful as group hunters than the local lions.

Spotted hyenas are robbers, too. They steal prey from other predators. A single hyena may chase a lone cheetah from its kill, or intimidate a jackal to leave its meal. A group of spotted hyenas can drive a leopard or even one or two lions from their quarry.

OPPOSITE Spotted hyenas tuck into a zebra carcass abandoned by a lone lion. Their cooperative bullying tactics (seen below) worked well.

ABOVE Spotted hyenas begin to circle a lion, in the hope that their threatening presence will intimidate the big cat into giving up its meal.

SUITED TO THE SURROUNDINGS

So which is the more successful or "better" predator – the black-footed ferret or the spotted hyena? It is not a very fair question, because each is suited to its ecology and way of life. If humans had not devastated prairie dogs and taken over the North American grasslands for farming, the ferrets would presumably still thrive. But the contrast between hyena and ferret shows how some predators are specialists, while others are generalists and readily "switch" their attentions between hunting a variety of small prey alone, pursuing larger quarry as a pack, or scavenging whenever the opportunity arises.

ABOVE Huge, fast and powerful – swordfish seem to be masters of the ocean, but like other top predators, their numbers are limited by food availability.

KEEPING PREDATORS IN CHECK

If hyenas are so adaptable and flexible in their approach, why do they not become even more successful? Could they increase in numbers at the expense of their rivals such as big cats and crocodiles?

The numbers of predators and prey in the African bush vary mainly with respect to the availability of suitable habitat and food supply. The predators do not have it all their own way. They are kept in check in many ways.

One factor that controls predator numbers is prey numbers. Large and powerful hunters like jaguars in the jungle and swordfish in the oceans may seem all-powerful. But they need a regular supply of large prey to sustain them. The prey themselves are often limited by the availability of their own food.

The process by which a herbivorous animal eats a plant, and is then eaten by a carnivorous animal, and so on, is known as a food chain. A shortage at the beginning of the chain, of the plants, has a knock-on effect along the chain. For example, a severe drought or series of wildfires on the grasslands of North America could affect the numbers of prairie dogs, which would in turn limit the numbers of black-footed ferrets.

TOO SMALL FOR SUSTENANCE

If a hyena runs short of large prey, it could switch to smaller prey such as mice, frogs and insects. However, to catch even a small victim, a hyena might have to stalk or chase it for a few minutes. The energy which the hyena uses up pursuing its meal, catching and chewing it, swallowing and digesting it, all adds up. It might become a total which is more than the amount of energy that the hyena obtains from its meal. In other words, the hyena could use up more energy itself in getting hold of its food, than it obtains from this food.

This principle of matching size of predator to bulk of prey is extremely widespread. Below a certain size limit, prey are simply not worth eating. They are not economical in terms of the predator's gain-and-loss energy balance. Even though the predator might eat often, it eats too little, and would slowly starve to death.

MORE CHECKS AND BALANCES

Predators which eat many kinds of prey feature in many food chains. They are the common links which draw the chains together to form a food web. The spotted hyena, with its adaptable approach, is part of such a multi-linked food web. If it cannot feast on its usual victims, could it switch to other sources of food?

To a point. But again, there are limits. Suppose that a serious drought has reduced the numbers of large prey animals. These are usually hunted not only by hyenas, but also by lions, leopards, jackals and other fairly large predators. As the prey animals decrease in number, there are more predators trying to catch them. The predators come into greater rivalry or competition with each other. This brings extra risks, not only from the hunt and being injured by the victim, but also from other predators as they fight between themselves to take over the meal.

RIGHT Jackals are smaller than big cats and hyenas. They usually band into extended family groups to drive a single rival from its meal. When times are hard, scarce prey means increased competition among predators.

WHY ARE PREDATORS RARER THAN PREY?

Quite simply, if there were equal numbers of predators and prey, the predators would have one meal and then starve. The reason that predators are much rarer than prey is that when a hunter eats its victim, it takes in some parts of the victim's body, digests some of what it swallows and its body uses some parts of what is digested. At each stage there are losses – what could be viewed as "wasted food". The food represents nutrients and contains the energy for life. The transfer of this food, and so its energy, from the prey's body into the predator's body, is never one hundred per cent efficient.

Similar "wastage" happens at each link in a food chain. Imagine a chain on the seashore. Seaweed is eaten by a young limpet. The limpet is eaten by a rock-pool fish, the blenny. As the tide comes in the blenny is eaten by a sea bass. At each of these links, from plant to herbivore, from herbivore to first carnivore, and from first carnivore to second carnivore, there are losses of nutrients and energy, as mentioned above.

Also, each animal in the chain uses up some of the energy in its food, as it carries out its own life processes, such as moving about, finding shelter, searching for a mate and breeding. This means the amount available to pass on to the next link in the chain, is reduced still further.

ABOVE Like other predators, tigers can never be more common than their prey because nutrients and energy are transferred upwards through the food chain, with losses at each step.

The energy equation

The overall result is that the food energy passing along a food chain decreases rapidly. At each stage, there are smaller amounts. The effect is known as a pyramid of energy. The pyramid is made of step-like, tapering layers. The plants form the base, which is the widest layer. The herbivores form the next layer. Their bodies contain or represent less energy than the plants, due to the losses and wastage. So their layer of the energy pyramid is smaller. Stage by stage, layer by layer, the energy reduces and can sustain fewer animals.

The peak of the pyramid represents the top predators. It has the smallest quantity of energy and supports an even smaller quantity of these top predators. The effect is greater if the animals are larger, as top predators usually are. A big creature needs much more energy than a small one, so a limited quantity of energy can supply fewer large animals compared to small ones.

This is why crocodiles and sharks may seem perfectly equipped to dominate their habitats – but they are always less common than their prey. It is a result of natural ecology, food chains and the pyramid of energy.

WARM BLOOD AFFECTS FOOD CHAINS

The effect of losing energy along a food chain is even more marked in food chains which involve warm-blooded animals – birds and mammals. A bird or mammal uses up great quantities of energy – in some cases more than half of its bodily supplies – simply keeping itself warm. This is in addition to the energy it needs for its other life processes. Which means there is even less available to pass on when a warm-blooded animal is eaten by a predator.

As a general consequence, food chains involving birds and mammals run out of energy more quickly. The pyramid of energy tapers to a point much more rapidly. So these food chains tend to be shorter, and result in fewer top predators. In theory, if lions ate only gazelles, and crocodiles ate only fish, then, other factors being equal, there would be more crocodiles than lions.

RIGHT Cold-blooded (rather than warm-blooded) hunters like the spectacled caiman – and the piranha it has caught – mean longer food chains with more top predators.

WHEN THE STAKES RISE

ABOVE For the cheetah, this encounter determines whether he gets his next meal. For the gazelle, the stakes couldn't be higher – life or death.

An encounter between a predator and its prey has very different meanings for each.

To the predator, an encounter is a chance to catch a meal. Unless times are very hard, it is not a matter of life and death. There will probably be another prey and another opportunity, later. Only a single meal is at stake. To the prey, an encounter is truly a matter of life or death. If it cannot escape, its existence is at an end.

The result is that prey are always under more pressure to "win" an encounter than predators are. The prey has more to gain if it can escape – it keeps its life. It is worth the risk of attempting to fight back or flee by any means possible. If the prey is injured or wounded in the encounter it might die later, but it would have died anyway if caught. The stakes can be no higher.

If the prey decides to battle, the predator might become injured or wounded. In the wild, wounds often become infected, and injuries rapidly turn into severe disability. This can easily threaten the predator's survival, since it might not be able to hunt. So as the stakes rise and the battle escalates, the prey has everything to gain, while the predator has much to lose.

THE ARMS RACE

ABOVE The jay is an adaptable feeder, consuming small creatures such as insects, and also plant matter like berries. When hunting, as a predator, the evolutionary "arms race" means that it lags slightly behind its prey in terms of equipment, methods and tactics.

This "life or dinner" principle, with the prey under more pressure to win than the predator, is one of the factors that drives the process of natural change that we call "evolution". It means that the defensive armour or escape techniques of prey are usually just a small step ahead of the hunting weapons and capturing methods of predators. Prey edge forwards, predators try to catch up, but prey stay slightly ahead in the "arms race".

Underwing moths are unusual in the moth group, since they fly by day, rather than hide away. However, they have excellent camouflage. With wings spread out flat, upper surfaces visible, they blend almost perfectly with the bark of a tree. This helps the moths to stay unnoticed by their main predators, which hunt by sight.

One of the chief enemies of the underwing moth is the jay. Some jays seem to learn how to spot moths by the faintly noticeable outline of their wings. So every now and then, a jay approaches a moth, with the intention of eating it. But the underwing moth has a second line of defence. It moves its front wings forwards suddenly to expose the bright colour on the upper surface of its back wings. The jay is startled by this flash of clear colour against the dull tree bark. In that moment of hesitation, the moth can fly away.

Do jays become used to this so-called startle coloration and learn to ignore it? Each different species of underwing moth has different startling colours, in shades of yellow, orange, red and crimson. An individual jay is extremely unlikely to encounter them all and learn to ignore them, let alone have the time and opportunity. The moths have taken the arms race a stage further, and maintain

8 PREDATORS AND PEOPLE

This book is mainly about the natural world. People feature here and there – mainly because we destroy predators, their prey, their habitats, or all three, usually for our own ends. But predators also have a special place in our minds and hearts. They are mostly big, powerful animals. We admire yet fear them, respect yet loathe them. Stories abound of people being attacked and eaten by lions, wolves, sharks and crocodiles, or being poisoned to death by snakes, spiders or jellyfish. Why should predators attack us, even when we are apparently posing no threat to them? And what is the real likelihood of a person being killed by a predatory wild animal?

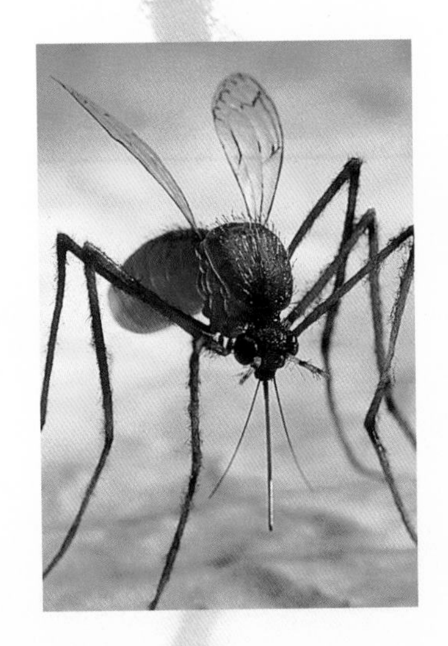

OPPOSITE The great white shark is one of the great predators we love to fear. Yet it probably presents a smaller risk, overall, than other lower-profile killers.

LEFT Blood-sucking insects, like this mosquito, are responsible for the spread of numerous killer diseases including malaria and yellow fever.

RISKS AND PROPORTIONS

It is difficult to estimate the exact risks of injury or death to humans from large wild predators. It is perhaps easier to put the risks into some kind of proportion. For example, in Africa, the big wild animal that causes the highest number of recorded human fatalities in an average year is not even a predator. It is the hippopotamus. Also, what about small animals? In all tropical regions, flies carry germs on their tiny feet, which cause immense suffering and death through transmitting diseases such as typhoid and cholera. Blood-sucking mosquitoes are responsible for spreading illnesses such as malaria and yellow fever, which kill more than one million people each year.

KNOWLEDGE AND PRECAUTIONS

For humans, some places are slightly more hazardous than others, because they harbour sizeable or venomous predators. There are lions in the African bush, tigers in the Indian jungle, poisonous snakes in the Australian outback, alligators in the American swamps, and sharks in many seas. These animals all pose risks, but for most of us, in our day-to-day lives, these risks are minimal and by taking a few basic precautions they practically vanish altogether. These precautions are simple and well known to people who are familiar with the ways of nature. Do not walk about in wild places at night. Do not stick your hands into holes. Wear stout shoes or boots rather than going barefooted. Do not approach or harass big creatures. It is mostly common sense.

THE PREDATORS' TOLL

Fatalities resulting from attacks by big predators such as tigers, leopards, bears, anacondas and sharks – even all added together – can get nowhere near the above numbers. Neither could they ever hope to compete with our own man-made threats to life, especially road accidents.

Not all predator attacks on people are confirmed or recorded. But in general, known deaths worldwide due to great white sharks average about 10 per year. Those caused by lions may reach 50 per year. The world total for deaths caused by crocodilians – crocodiles and alligators, and their close cousins the gharials and caimans – probably runs to more than 1,500 each year. It may be as high as 3,000. That is almost 10 people per day. However, you are thousands of times more likely to meet your end when crossing the street in your nearest town – the World Health Organisation estimates that 2,700 people worldwide are killed every day in road accidents!

LEFT AND OPPOSITE
Great whites are responsible for occasional fearful injuries to people. But road accidents produce equally terrible wounds, and in much greater numbers.

BELOW Jellyfish lack speed, claws and teeth. They are hardly even visible. Yet they are effective predators with powerful stinging chemicals.

SELF-DEFENCE

Why do small but venomous predators such as spiders and snakes bite people? They could never hope to make a meal of us. Often it is an entirely natural response – self-defence. The poison is designed both to subdue prey and deter predators. Surprised by a huge and clumsy human who had just blundered into its web, or trodden on its tail, the spider or snake defends itself in a way that has been shaped by evolution, over thousands or millions of years. It strikes back in a reflex action to aid its own survival. The same usually applies to other small but poisonous predators who allegedly "attack" people, such as the blue-ringed octopus, coneshell or scorpion.

Often a venomous creature precedes its strike by a threat display designed to warn off the enemy. It may even carry out a preliminary strike, biting or stinging, but not with its full power. Many larger animals, finding themselves perceived as a threat by a little but dangerous adversary, respond in the natural fashion. They recognize the warning signs and move away with caution. Many people do not. They are curious to see how far they can "push" the predator, or they are eager to show their domination of nature. When the real venomous strike comes, it happens too fast to avoid. The person suffers the consequences.

NO HARM INTENDED

It is probable that predators such as the stinging jellyfish and box jellies (see page 57) either sting human swimmers passively or in a self-defence reflex. They do not distinguish between humans and other objects or animals in their path. They simply trail their tentacles through the ocean, as they have been doing for hundreds of millions of years.

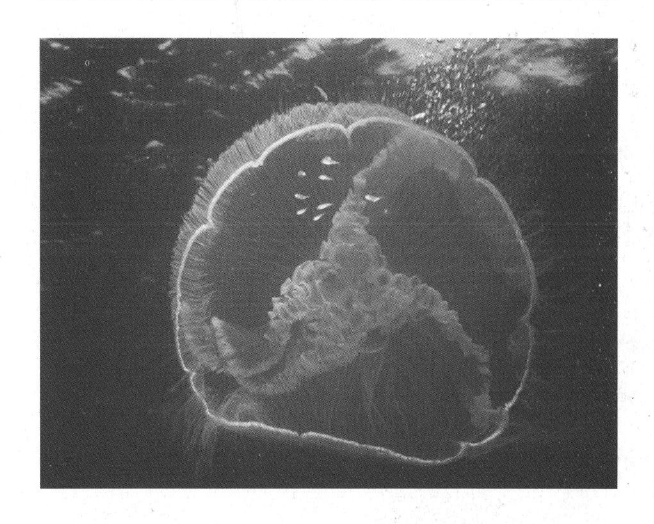

SURPRISE ENCOUNTERS

Many larger predators that supposedly attack people are also acting in self-defence. Out swimming, a person may suddenly come face to face with a barracuda or shark. In the woods, it might be a bear or big cat. In a swamp, it could be an anaconda or crocodile. In a rocky, remote place, it could be a wolf or a mountain lion.

Who is most surprised? Who is afraid of whom? Many creatures have a set of in-built behaviours, or instincts, which they carry out rapidly and automatically in response to a sudden event in their surroundings. Such instincts have been modified and honed by evolution to cope with events that crop up in their lives now and then. But a human being wandering about in an animal's own habitat is usually a new event for the animal. Often, it has no experience to prepare it for this encounter, and no instinct that it can apply to the situation. So it "panics".

The human is in the same predicament and may well panic too, with screams and shouts, arms waving and legs kicking. This may serve to scare the predator off – animals are programmed to survive and most use avoidance as their first defensive strategy when threatened. However, the human may be unlucky and their reaction may prompt the predator to go for "the best form of defence is attack" approach – and quickly. Whether it pursues "flight" or "fight", the predator is following its natural tried and tested pattern of hunting behaviour in order to protect its own life. A person, on the other hand, is more likely to pause, try to weigh up the situation, and think through the options. By then, it may be too late.

BELOW The anaconda has been accused of many attacks on people, and even of swallowing humans whole. But then, its size has also been exaggerated, with claims of sightings of individuals measuring more than 50 metres for a snake that rarely exceeds 8–9 metres in length.

ABOVE To us, this is an underside view of a swimmer paddling on a board. But sharks know little of swimmers and boards, only of general seal-like shapes that could mean food.

SPECIAL CIRCUMSTANCES

Some circumstances make attack by the predator more likely. It could feel cornered, with no chance of escape, so it tries to fight its way out. It may be wounded or diseased and in pain. It may be a parent defending its young – often one of the most dangerous situations. It may even mistake the human for one of its regular prey. When seen from below, a person lying on a surf board, arms paddling and feet trailing behind, forms a silhouette which vaguely resembles a swimming seal. Great white sharks like seals.

Very occasionally, a large predator deliberately hunts a human as prey, if the opportunity arises. It has happened with the larger cats such as tigers, lions, leopards, jaguars and pumas (mountain lions or cougars). It is also said to occur with wolves, coyotes, dingoes and other pack-hunting members of the dog family, some kinds of bears, and of course predatory fish such as the larger sharks. It is not so much a deliberate campaign to catch, kill and eat a human being, but part of the flexible, adaptable behaviour that many predators show. They are ready to take advantage of any victim which might come along.

In a few cases, especially with tigers and other big cats, it seems that predators deliberately choose to hunt humans, in preference to their natural prey. This is when they truly become "man-eaters". Often they are old or disabled individuals, which can no longer catch their usual victims. Their human victims tend to have no experience of a sudden struggle for life and death. People are soft targets.

THE OCCASIONAL PREDATOR

The chimpanzee is an example of a seemingly peaceful animal that can transform into a dangerous predator. Most chimps live reasonably peaceful lives in the woods and forests of West and Central Africa. About nine-tenths of their diet is vegetarian, chiefly fruits and also soft leaves, buds, blossom, sap, roots and bulbs. However, chimps also eat animal matter, especially termites, and eggs and baby birds. Occasionally several male chimps from a troop form a hunting party that sets off to find more sizeable prey – a red colobus, young baboon or similar monkey, or a small deer.

The chimpanzee hunting party may spend several hours on the trail of its quarry. It attacks in a coordinated way, surrounding the victim and blocking its escape routes. The members whoop and clamour with enthusiasm as they rally together. When the prey is caught, the chimps bite and tear it apart, and wave the pieces around, apparently screeching in triumph. They may share the meat with other members of their troop, including females. The success rate for some of these predatory chimp parties is four catches from every five attempts.

BELOW A chimp displays its large canine teeth in threat. This great ape is capable of bloodthirsty attacks, even on its own kind.

Why does a group of quiet chimps suddenly turn into a terrifying pack of fierce hunters? There are several possibilities:

- One suggestion is that the meat provides valuable proteins and other nutrients that are lacking in the usual vegetarian diet. But chimps still hunt, even when the forest is full of the most nutritious fruits and other plant matter.
- Another idea is that a male chimp uses the meat, which is a prized and tasty snack, as a "present" to attract females for breeding. However, studies show that many males mate about the same number of times when they give meat to females as when they do not.
- A newer theory is that the hunt helps to strengthen the bonds or "friendships" between males. The male chimps in the party who share the meat with each other are also more likely to do other activities together. They groom each other more, and aid each other in battles for a higher rank in the troop. The hunt may not be for nutritious food, or to gain favours with females, but to reinforce male bonding.

ABOVE Like a child in a sweetshop, the fox in the chicken coop is overwhelmed. It may kill many hens, and return later to retrieve or store the excess. But its natural instincts are not designed to deal with such unnatural concentrations of food.

NATURE'S REVENGE

Predators have suffered a catalogue of persecution and destruction at our hands. Through history they have been trapped, shot, stabbed and poisoned using an ingenious array of methods. Why? Because some people consider that killing a powerful wild animal – even from hundreds of metres away with a high-powered rifle – shows bravery and provides a trophy to hang on the wall. Because killing and dominating dangerous predators offers a psychological way for man to overcome his fear of them. Because predators represent a threat, either real or imagined. Because occasionally they attack our farm animals or pets. Because very, very rarely, they attack us.

In ancient times, when people lived closer to nature in caves, large predatory animals were a real threat. But does this apply in the modern world? Humans are so successful as killers that we have almost wiped out many large predators. Tigers, great white sharks and many others are on the lists of threatened species. Given the way that predators have suffered at our hands, they might seem to have every right to get their own back. Luckily for us, animals do not appear to bear grudges or seek revenge against humans, in the way that we do against them – they are simply attempting to survive in the world as they find it.

Predator profile: Tiger

Dusk deepens. Monkeys fall silent as insects chorus with buzzing chirps. After a hot, steamy day, the tiger rises from a cool, muddy pool. It slinks silently down a bank into a small stream valley, one of its regular patrol routes. After an hour's stealthy prowling, with several false trails, a tiny noise catches the cat's ears. Creeping closer, it peers through scrub to make out a lone male sambar, a large and powerful deer, munching soft leaves. Luckily for the predator, the prey is upwind and so less likely to detect the cat by scent or sound. The tiger slowly lowers itself into a semi-crouch and begins a long, cautious stalk. It must try to get within charging range, 20 metres or less. The sambar sniffs the air and looks around occasionally. But the cat freezes instantly and relies on its striped coat in the dappled moonlight. This could be the fifth night without a meal – tigers are successful in fewer than one in ten hunts. But the tired sambar is too busy browsing, hungry after a day battling with rival males for territory. The tiger steadies, tenses and springs forward. After four massive bounds it hurls itself at the victim ...

Predator profile: African wild dogs

The sun has not reached the horizon when the African wild dogs begin to pace near their den. Steps quicken as they yap and snap. It's motivational behaviour, psyching up for the hunt. Then the pack of six go quiet and trot off as one through the dimly-lit bush.They fan out and pause often to gaze, sniff with their long muzzles and prick up their huge ears. Suddenly one dog near the end of the line accelerates into a more urgent canter. The others respond as they discern a small group of impala through the dawn haze. The alert quarry burst into a sprint. The lead dogs see immediately that one impala has a wounded leg. The slim-bodied, long-legged hunters change gear again, seeming to cruise effortlessly through the bush, yet notching speeds close to 60 kilometres per hour. The injured impala runs gamely, but the dogs gain. The pack outriders begin to circle into a "C" formation. The victim changes tactic and tries to dodge, but finds itself with dogs front and rear. One leaps to grab its nostrils while the other hangs on to its tail ...

Predator profile: Jackson's chameleon

In East Africa, a green "leaf" slides along a branch. It's a lizard – Jackson's chameleon. Like all chameleons, it can change skin colour and pattern to match its surroundings. Such detailed camouflage brings a double benefit. The chameleon stays unnoticed by prey and is less visible to hunters who like lizards for lunch. In practice the chameleon has two main colour modes: green for fresh leaves or mossy, lichen-encrusted bark, and brown for old leaves and bare bark. This is a green day and the slow-motion hunter grips its branch strongly with opposable toes. It could hang upside-down or use its coiled tail as a fifth leg to help it remain stable and move steadily, as it sneaks at snail's pace towards a fly. The chameleon's turret-like eyes swivel to look in every direction separately, and finally lock on to the target. It can only score a direct hit if its vision is spot-on. Just another few millimetres and the chameleon's tongue shoots out, longer than its body, and too fast for our eyes to follow. Its saliva-sticky tip slaps on to the fly, the tongue and prey are flicked back into the chameleon's mouth ...

Predator profile: Great white shark

The seals are reluctant to leave the beach. Yesterday a carcass washed up – the front half of one of their group. But hunger mounts, and they cautiously enter the water. Watching for shapes or shadows they swim deeper. Suddenly a massive shape surges from below, smashes one seal clear of the water, and immediately disappears again. The great white shark has retreated after its single fierce bite, to avoid a fight which could injure its precious eyes and nose. The shark has lurked for hours, attracted from more than ten kilometres away by the scent of blood, when one of its kind feasted yesterday. From a depth of 20 metres, its eyes spot a likely silhouette against the bright water's surface. The seal sees little, since the shark's dark grey back blends into the dim water beneath. On its first attacking burst, the shark's one tonne rams into the prey at 50 kilometres per hour. The fish's rear half is nearly all muscles, "superheated" by a special blood system to work extra-fast. Just before impact, the shark rolls its eyes into its skull for protection, and relies on electro-sensors around its face to sense the seal's active muscles. By now the prey is dead, and the great white moves in leisurely to feed.

Predator profile: Sydney funnelweb

Dusk turns to night as the funnelweb finishes spinning its long strands of silk. They are like guy ropes extending from a funnel-shaped web, that leads to a nearby silk-lined hole. The spider retreats into this lair, and waits and watches. When its delicate sense of touch feels a victim moving on the threads, the spider rushes out and attacks swiftly, surprising its prey. It has relatively enormous chelicerae, or fangs, for its body size, and these point almost straight downwards. But the fangs and the associated mouthparts can hardly move in relation to the spider's whole head. So, to impale the victim – tonight it's a leaf-hopper – the spider must rear up and then strike downwards with its whole body, stabbing the fangs into the prey. As its front legs hold the prey, the fangs repeatedly inject drops of deadly poison. This is powerful enough, not only to subdue a mouse or small lizard, but even to kill a human. This is why, in the funnelweb regions of south-east Australia, people take careful precautions to avoid a possible brush with death. They wear protective clothing, gloves and boots. Funnelwebs love dark crevices, so never put on those old gardening shoes without shaking them out first ...